AF573730

Airedale & Wharfedale College
T07417

THE GREAT SKIPTON SHOW

By the same author

GRANNY'S VILLAGE

The Great Skipton Show

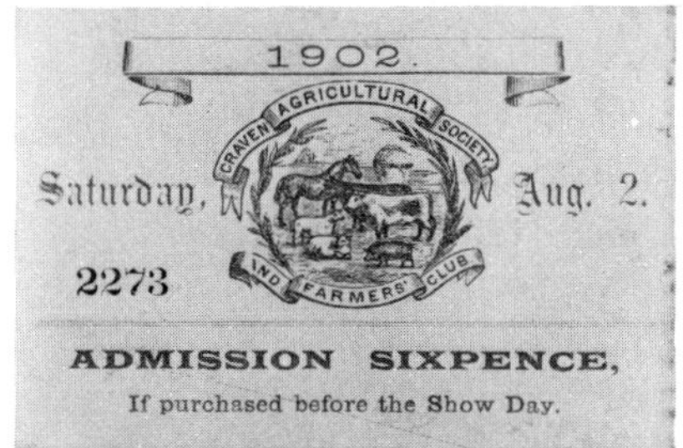

ROGER MASON

PETER DAVIES : LONDON

Peter Davies Ltd
10 Upper Grosvenor Street, London W1X 9PA

LONDON MELBOURNE TORONTO
JOHANNESBURG AUCKLAND

First published in Great Britain 1979

432 09211 0

Printed and bound in Great Britain by
Fakenham Press Ltd, Fakenham, Norfolk

CONTENTS

LIST OF ILLUSTRATIONS

For
Roger Parkinson

ACKNOWLEDGEMENTS

I would like to acknowledge the help of Dr G. Rowley, the author of 'Old Skipton', and Mr B. Wilkinson and the Skipton Local History Society; Miss Harding of the Craven Museum; Mrs K. B. Robinson and Skipton Civic Society; Mr Arnold Fell and Mr Stockdale, the farmer (who is no relation to the Stockdales in this book); Mr Harry Fell, of the former Fell's Craven Leadworks; Mr Ian Plant and the *Craven Herald;* Miss Fairey; Mr Albert Toy, formerly butler at Skipton Castle; Wilby Padget; Edgar Leach; and especially my aunt and uncle, Alice and Jim Tosney, not forgetting his aunts, Belle and Annie Tosney. Between them they have given me a surprising amount of varied information and ideas, perhaps without always realizing how it would be used. I hope they will forgive me and enjoy the book.

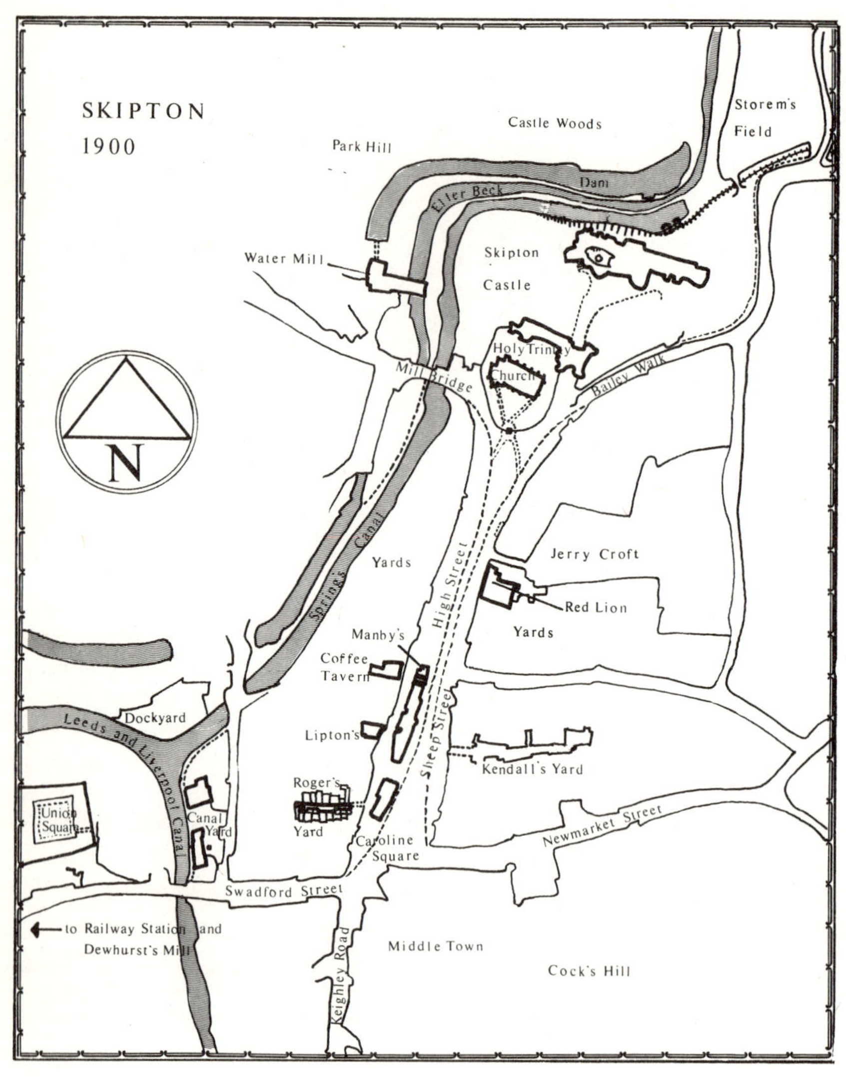

Skipton at the turn of the century

I

IN THE DALES

The eve of Skipton Show was a hot, still day, promising well for the festival, a day that flaunted its glory from a brassy, glaring sky, one that drove Simon Stockdale out at sharp dawn high on to the backside of Malham Moor. He lived far up in those green limestone dales of Craven cut into the Pennine hills to the north of Skipton, a steep landscape of stone villages and remote farms whose folk sent their produce to market in the valley town; villages and farms where all the plans for Show-time had been long made and interminably discussed. Sweating farmers had been shaping up future prizewinners, cattle, pigs, poultry and above all sheep, in the dusty grass of late summer fields. Tomorrow, the show Saturday, would be a full and busy day, one of the great days of the year. Haymaking was over, and well over this time; winter preparations, pig-killing and the like, yet to come. Jasper was out, chewing the cud complacently in slanting sunlight, as Simon passed him on the way up to the moor and wished the Lord's blessing on him.

The Skipton Show, after fifty years or more as the annual event of the Craven Agricultural Society, was now a well-established tradition. It was flourishing at the beginning of the Edwardian Era, shortly after the old Queen's funeral, when Jasper was in his second year, when the young lime trees planted along Skipton High Street for the Queen's Diamond

Jubilee were growing into bushy saplings, and when Sally, Simon's granddaughter, was stirring dreamily in her mother's old bedroom at the back of the farmhouse, with its stone-mullioned window looking down into Littondale and the morning silver of the little river Skirfare. Those lime trees would give scant shade next day to Craven farmers come down into town from every dale, from Airedale, Wharfedale, Ribblesdale, all the names that echoed the story of a Saxon and Scandinavian past, of a long-settled pattern of life with its own peculiar ways, customs and sayings.

In the Stockdales' own Littondale too, in that farm high out of the valley, the morning's work had gone steadily onwards as Simon finished his two-hour stint of peat cutting, and breakfast time was at hand. Martha Stockdale stood in the sunlight at the open kitchen door. The room behind her was an etching in black and white, of heavy beams set against lime-washed walls and ceiling, dark oak settle, table and ladder-backed chairs, the pale sunlight reflected in polished steel knobs on the kitchen range, a sombre, leaded enclosure for the glowing, crumbly, summer fire. 'Well,' she said, 'here I stand dreaming when there's work to do,' so she turned towards the bakestone which had been glowing with the heat of a peat fire since six o'clock. 'Should be good and ready,' she muttered and discussed the issue briefly with herself as she laid out the baking implements. She had seen her husband's broad, short-necked figure in the field above the shippon, coming home, spade on shoulder, from the moor. Downwards he strode, still a good walker despite his nearly sixty years, with a typical, energetic bounce upward from the balls of his feet. He vanished behind the farm buildings and Martha shook herself, saying out loud, 'I'll have to get things fettled sooner . . . these hot mornings don't suit at all, not at all.' A few moments later, as she riddled the fire under the bakestone, and a bead of sweat ran down her nose, she added, 'not at all,' with emphasis.

'Talking to herself again . . .' the gruff voice of her husband rang through the room. 'Oh, Simon! Tha made me jump. Why didst creep up on me like that!' She walked over to where he stood, his mouth held tight, jawbones working to contain a threatening outburst of laughter. Martha's clogs squeaked on the sand newly spread over the flagged floor. 'Now take that silly grin off thy face and come in and sit down. I've put t' kettle on and there'll be fresh riddlebread for breakfast, see, there's t'batter rising. I'll make a good rack full,' she said, half to herself, half to her husband, 'then it'll last us well into next week.' Simon rubbed his hands in anticipation of hot, fresh oatmeal cakes for breakfast. ''Twere good meal, that last batch,' he said, 'still had the taste of fire in it.' The best oatmeal came from fire dried oats, slightly parched in the drying. 'Aye,' retorted Martha, 'and you've tramped it down that hard in the ark with your big hooves I had to take a chopper to it to get it out.' 'You've got to keep the air out,' grunted Simon in self-defence.

Martha stirred the batter and then poured a ladleful hissing on to the bakestone. She ran a flat scraper over the top, squeezing the cake to its ideal thickness. Steam began to fill the kitchen and the mealy, porridge smell of cooking. Deftly she lifted each thin, roughly oval shape from the stone and hung it to keep on the slatted wooden flake above. Like a row of tattered grey washleathers, the flaps of bread swayed under the whitewashed ceiling. Simon stretched out his legs for a moment of quiet rest, drying the dew from his stockings at the fire.

'Sardy, Emerald, Beryl, get on wi' it . . . budge, can't yer!' The high despairing tone of Joady Thacker's voice jarred through the silence in the kitchen, a voice that hissed with irritation as the skittish cows invaded the yard, taking a quick bite as they passed at Martha's border nasturtiums. 'What's Joady up to?' said Simon peacefully, opening his eyes, but reluctant to leave the fireside. A long, enquiring face pushed in through the doorway, eyeing him solemnly, with a string of orange and green nasturtiums fluttering from the yellow teeth.

'Pearl – get out with you,' shouted Martha, good-humouredly, as she shoved at the cow's nose with the heel of her hand. 'Joady,' she called, 'Joady, whatever are you doing . . . ?' But Joady was back now with his thumbstick, which he always left behind one of the boskins, the stone partitions in the shippon. He shambled to the yard gate, blinking as if every step was an adventure into the unknown, and waved his stick. 'I'd best go sort him out!' Simon grunted, slapped his knees and went. So Sardy, Emerald, Beryl, Amethyst and Pearl, convinced by a show of authority, followed Joady complacently towards the low pasture, their bit of fun behind them. Joady smiled to himself. They were good beasts at heart, no malice in them.

Martha Stockdale shook her head in wry amusement at her husband every time she thought of the names he gave his beasts. Admittedly his favourite reading in the Bible was the Book of Revelation, but was it proper to use Holy Scripture for naming cattle? 'The first foundation was jasper; the second, sapphire; the third, a chalcedony; the fourth an emerald. . . .' Of course, Joady shortened sardonyx to 'Sardy' and he found chalcedony totally impossible, but otherwise Simon worked to a regular system and the names of their cows followed each other in her mind's eye. There was Beryl the first, short-legged and a good milker, Topaz who always kicked over coppy and pail at milking time if she could, Jacinth, that Joady always called Jacky, and who had been such a pretty calf.

As if in response to her musings, Jasper the bull bellowed over the wall at the passing cows from his small croft behind the shippon. Joady had given him a good scrubbing and polished his horns yesterday but he was by no means ready for the Show. Martha smiled to herself at the thought of Jasper. They had raised him by hand and he was always good-natured, like one of the family. A beautiful bull was Jasper, well put together, not too massive but muscular under silky skin, fit to be cast in bronze. Cleaned down now and brushed, his markings, red and white dappled roan, with darker ox-blood patches here and there, shone in the sunlight and his bellowings had an air of boastful well-being.

As she methodically smoothed the riddlebread, piled the flat soft flaps or hung them on the flake, Martha thought over the year. Now it was slack time at the farm, haymaking over, hay safely stored in the outlying laithes, hay sledges cleared away until next year. Some of the children had come back home to help. It was needed at haytime, for, though Simon could mow as well as any, Joady was a hopeless, gangling confusion when handed a five-foot scythe and at his best could manage half an acre a day. The children had gone back to their regular jobs in the towns, but at least they'd left young Sally behind as company for her Grandma. Growing pains of course, nothing more, and to be expected at that age, with a lass. You couldn't grow into a woman without some discomfort. So here was Sally – her Dad an overlooker in a textile mill – staying on with her Grandma for the good dales air. She'd soon fatten up and lose that pale face. Though at times she seemed a quiet, self-contained, quaint child and a bit young for her age.

Martha wiped her forehead. Yes, it was slack time at the farm and the men would be wanting a good long spell over breakfast. More work for the womenfolk. The house felt empty now that the grown-up children had all gone, empty and echoing with memories during the times when she was left there alone. Yet Martha chuckled when she thought how they had taken on Joady Thacker to help. He must have been over fifty, ten years ago, when Simon hired him, in his baggy breeches with a grey plaid wrapped round his shoulders. Unmarried and shy he was, all his life a cowman. They had room and to spare in the house, but Joady had doggedly, silently, looked downwards at the suggestion. 'No,' he whispered, 'It bain't right. I'd best be near t' beasts,' and, as he shuffled his feet, he explained how he must sleep above the cattle, how he could feel what they were thinking and how he'd gotten that used to it that he could never sleep without the soft grunts and stirrings in the night and the sweet smell of grass-fed breath.

A shadow darkened the kitchen as Simon returned. Martha straightened her back. 'I've done three dozen,' she said, 'and I know it's time for breakfast, so there's no need to speak. Will you call our Sally in and see if you can get Joady up from his cows in time . . . !'

Simon stumped outside, his broad-cheeked face drawing up into a smile at the thought of his grand-daughter. Children always found the best places and he knew where she would be, where his own girls used to go to play in quiet. The dairy stood at the end of the house, before the shippon; it was an outshut, a lean-to with sloping, stone-tiled roof that sat retiringly back from the main part of the house and had its own little flagged yard before it. In the narrow gap between shippon and dairy grew an old, bent, rowan tree, its berries just turning from green to scarlet. Back against the dairy wall, under the window, was an elder bush where Martha would spread out her butter muslin and drying cloths in the sunshine.

These two trees defined a sheltered area of grass, facing south and backed by the high, rough shippon wall. A stone-lined cistern was set in the back at the edge of the flagged yard, full of limestone water so clear that it seemed to bring every pebble in the bottom up to meet the child's eye that sought it. Spring water trickled constantly into the back of the pool and out over the green edge at the front, to fall underground amidst mosses and a clump of ferns. This cistern supplied buckets of water for both shippon and dairy, being handy for either.

There could be no more delightful sanctuary for a child, and Simon found his grand-daughter sitting in sunlight under the rowan tree, head bent, intent on her game. He squatted down beside her. 'What's tha doing, lass?' Sally looked rather shy at being suddenly dragged out of a secret game into ordinary life and perhaps a little embarrassed at her childish pursuit, but Simon had a confiding serious kindness bout him, a grandfatherliness that was irresistible. 'It's a farm, grandad,' she said. 'Look, there's the fields,' and showed him the scratched outlines, the dividing walls made up of pieces of twig, and the grassy hillock that represented the high fells. 'And what are these?' He pointed at the collection of big and little fir cones huddled in flocks around the sloping fields that ran down to the cistern.

'They're sheep and lambs, of course.' The young voice was rather sharp in surprise at his ignorance. Simon grinned to himself. 'Takes after her Grandma,' he thought. 'And this one?' A large, rather battered fir-cone was in a field of its own. He picked it up. A length of stalk dangled from the end. 'I see,' he added slowly, 'it's the old Tup, himself,' He flicked the length of stalk with his fingers and, chuckling, laid the tup back in its place. Sally giggled. 'Come on, lass, it's time for breakfast.' Simon held out a strong hand and she took it confdently, leaving her game, ready to help her Grandma now that play was over.

The table was laid already, teapot, mugs, jug of milk, plates, a mound of yellow butter and a grey heap of hot riddlebread still steaming, whilst as a real treat, an extra brought out in honour of Sally, a tin of treacle stood beside the riddlebread. 'Where's Joady?' asked Martha. 'Haven't seen him,' replied her husband. 'Well we'd best begin. He'll come when he comes.' She laid a soft flap of riddlebread on Sally's plate, rough side up, and spread butter on it with a lavish hand. 'Help thyself to treacle, lass,' she said, taking her own chair at the table-end by the fire.

They ate and drank in peace as the sunlight shone steadily on the fells that rose beyond the farmhouse. The fog, the fresh new grass that springs up after haymaking, brightened intake fields at the edge of the open moorland and they could hear the distant skylarks. The warmth and stillness seemed to increase their calm. 'Are t' ready for the Show, love?' said Martha to Sally. 'Yes, Grandma. I'm going to wear my pink dress. . . . It's all ready ironed. And can I borrow that linen bonnet, from the cupboard?' 'Aye, of course you can, love,' Martha smiled in memory, 'it were thy mother's, and right pretty she looked. . . . We'll find some ribbons.' 'She'll be a pretty young rabbit herself,' said Simon, stroking her fair, wispy hair as it dropped down beside her face like long silken ears. Sally brushed it aside with a somewhat treacly hand and smiled enticingly at him. 'Are you going to buy me some ribbons, Grandad?' she asked with a mischievously seductive air. Both grandparents laughed. 'Nay, wait and see,' said Martha, as Simon chucked her under the chin in mock reproof, 'pass me t' *Craven Herald,* Simon.'

She pored closely over the Show advertisements in the newspaper's narrow columns. 'What are you putting Jasper in for?' 'Best short-horn bull under eighteen months,' replied Simon promptly as he brought his chair round beside her, 'dost remember how well his father did?' 'How did you like that last lot of butter?' Martha interrupted. 'It were very sweet.' 'Aye,' mused Martha. 'I don't reckon I've ever made better butter. It's Beryl, you know.' Simon raised one eyebrow in reply. 'Her milk's that rich, even off fog. . . . Mind you, it's slow churning.' Martha rubbed her hands together for a moment, as if summoning energy for more work. 'Listen to this.' Simon raised his bushy eyebrow again, '"Special prize, donated by Mr Walter Morrison Esquire J.P. Best basket of butter being not less than two pounds to have no distinctive mark or print on it, to be made up in one pound pats absolutely without salt. First prize five pounds." That's a right good sum, but then, we know Mr Morrison. He'll not miss it. You'd have to make it today, though.' 'Aye, I know that, you . . . you husband! Any road, if it's not to be salted it wants to be fresh to keep sweet. . . . Ah, here is Joady, late as usual. Don't you like my cooking young man?'

Joady was blowing a little after his brisk walk up from the bottom field. He rested his thumbstick carefully against the wall just outside the door, wiped his face with his hands and his hands on his breeches and came cautiously over the threshold. 'Warm day . . .' he said softly as he slid his glance rapidly over the room. His eyes were still a startling blue in the creased and weather-beaten face. 'Can I help, Grandma?' Sally had swung from her chair and was making for the door. 'No, you go and play, love, you're on holiday,' answered Martha. Sally stopped before Joady's scarecrow figure and smiled up into his face. To her he seemed more like some friendly animal – a big dog perhaps – than a real adult who gave orders and took thought for the morrow. 'God bless ye, missy,' whispered Joady, and as he shuffled a cautious little jig to her, his accustomed, timid half-smile broke into a heart-warming grin that lit up his whole face. Sally, anticipating talk, decided to sit down on a stool under the window and watch them eat.

'Come along, Joady,' said Martha, 'sit down and take some breakfast. I've made hot riddlebread . . . sit ye down.' Joady slid into a chair and looked across at Simon before beginning. 'Smith's coming,' he said. 'Aye. Good.' said Simon. 'Get him to shoe Jasper before you set off for Skipton. Ten miles'll do no good to his feet.' 'Am I still going?' Joady's words were barely audible. 'Of course you're going,' said Simon in exasperation, 'don't be so soft, man. Skipton's not like walking to London. Aye, and you'll have company. . . .' Joady looked up from the table in surprise, 'Jasper'll be right taken with the trip. He'll tell thee what he thinks of the big city . . . once you're landed safe.'

'I've been thinking things over,' Simon spoke directly at Martha and she, knowing the tone, gave him her full attention. 'We've had a good year. Few lambs lost, good grass early and then hay that dry it were almost blue. There's one or two bits and pieces we could do with getting if we had a sovereign or two to spare. We can live another year. I'm not worriting, mind, so don't you grin like that.' Martha shook her head and eased the smile away. She knew well enough how hard things could be on a small stock and sheep farm. They were never easy and so it gave her pleasure to hear Simon actually talking of spending money on 'bits and pieces'.

'I've a couple of good Wensleydale lambs coming on,' said Simon, still seriously pursuing his thoughts. 'They might stand a chance at t'Show and it's daft not to take all ye can if ye're going any road. We'll sell 'em whatever happens and have some brass to spend. Though I don't want you womenfolk to go mad.' This was not to be taken seriously and they all knew it. 'I do remember,' Martha commented in a subdued voice, 'the times we hadn't a single beast worth putting in and we couldn't even bear to part with a couple of shillings for entrance money, so we stayed at home. . . . Now Jasper should do well.'

'Aye, I've great hopes for Jasper, and Joady's more work to do on him.' Simon rubbed one hardened hand over his cheeks. 'If he wins I'll keep him on to serve heifers. If not, then he must go; once he's got to Skipton there's no point in bringing him home.' This remark brought a moment of gloom to the kitchen, but no comment. Simon shrugged his shoulders and held out a cup for more tea, adding, 'Joady'd better let him know he wants to win!'

The kitchen was silent for a while, except for the sounds of eating, until the noise of clogs came across the yard and a small sinewy man peered in through the doorway. 'Morning

Simon, Mornin' Mrs Stockdale. . . . Joady.' 'Morning John,' 'It were just them half-shoes for Jasper, weren't it?' 'Aye, but come in and have a sup of tea.' 'Right you are. I'll just drop this lot.' There was a sound of rattling irons and then the blacksmith came inside.

Martha brought him over a mug of tea and pushed the shrunken heap of riddlebread towards him. 'Nay, I've had breakfast,' said John, 'and I've a tale for thee.' 'Th'art a good one for a tale, John,' said Martha. 'Aye, perhaps. It were along of being down at William Wilkinson's, fitting some chains on a hay rake.' Simon nodded, it would be a horse-drawn rake needing new links on the towing chains. John continued. 'I saw them beasts he bought off Jemmy Rattle in Skipton.' 'Aye, I reckon they'd look mighty pretty. Did he pay a good price for 'em?'

'Wouldn't tell,' said John, making a wry face, and they all laughed. Jemmy Rattle was famous as a dealer in cattle, who could beautify the scrawniest old Scottish cow into a five or ten pound profit over what he paid for her. 'Aye,' said John, 'but I've looked 'em over, and I had a quick word with the lad who brought 'em. I know a few tricks of the trade now,' and he winked slyly at Simon, 'after he'd had threepence for a glass of ale. . . .

'First he washes 'em down well and rubs their coat up wi' sawdust. Gets a good gloss on you know, so's you can see the shine of a new sovereign against 'em! Bit of bloom you might say, on an old plum. Then he saws tips off the horns and chisels a good rounded end to 'em. The ladsays folk like a nice smooth horn. Then he wants to polish 'em to a good shine, so, being a careful man, he puts his fingers in their ears, get out some wax and works up to a right smart finish.' Martha made a wry face. 'I don't like that Jemmy Rattle,' she said, 'he's got a greedy face, a black face. . . .' She turned to Simon, hands on hips, 'I don't like you selling beasts to him. He's too clever by half, and he doesn't think they're living creatures, the Lord's creation . . . Folk shouldn't deal with the likes of him.' Sally, who had been listening from her stool, shuddered at the sudden vision of a leering devilish face gloating over the corpses of her friends the cows.

But the tale was not yet finished. 'Oh, you listen a while,' said John, 'there's worse to come. Of course, he gives their hooves a good polish an' all, but the real trick comes in at the end, the day before market. Then he goes round wi' a bucket of collodion, isinglass you know, and he dips all the teats in it. Just the ends, not so's owt would show.' 'The old beggar,' started Martha indignantly. 'That seals 'em up till it's washed away, and so the bags swell to a fair old size. . . . And the last little touch,' John's hand was held up to forestall Martha's comments, 'just as they're about to go out, like womenfolk in general, he ruddles the bags up well with a dusting of rouge. . . . I told you he makes 'em pretty.' Simon whistled. 'Well I never,' said Martha in disgust, 'and he gets away with it!'

'Seems to,' said John, 'or else William Wilkinson can't resist a pretty bag.' Simon laughed, Martha grunted and Joady rose to his feet, making for the door. 'I'll come wi' thee, Joady,' said John, 'just get Jasper into t'shippon and we'll have him shod in a jiffy.' Simon went with them. Martha and Sally were left behind to clear the table.

When that was finished Martha walked round the corner to the cool, whitewashed dairy, its wall flickering with sunlight reflected from the trickle of water that fell into the cistern, inside a haven of calm and unhurried movement. She looked at the cheeses still maturing on the slatted shelves at the back. 'They're still running wet,' she thought, 'it's going to be hot again.' Last night's surplus milk had been set to separate in a long shallow, lead-lined tray. Now she turned the tap below and skim milk trickled thinly into the bucket below. The thick cream was left adhering fattily to the lead, ready for Martha to scrape it briskly into a bowl with the horn scraper.

'Be about a gallon,' she thought, 'say three pounds of butter. Now where did I put that cream from yesterday? I'll make it up to a full two gallons.' She had a two-gallon tumble churn, from Waide's of Leeds, a big improvement after the old plunger churn they had used until a couple of years ago with its backaching pumping motion. Now she set to work, turning the handle with a will, and the cream sloshed rhythmically to and fro in the churn. The morning was slipping away in the old familiar pattern. As she churned Martha had time both to look out of the window and to survey her tidy dairy, noticing the old witchstone, a perforated piece of limestone shaped like a rough star, water-worn, with pock-marks and smooth indentations, that had been her aunt's, and now hung by the separating tray. It was supposed to keep witches and their souring influence off the buttermaking. She could feel the butter gathering itself together as she churned.

The angle of the sun changed slowly on the dairy floor. Cheerful shouts of the men could be heard as they emerged from the shippon. Joady was leading the bull by his nose ring. Jasper seemed slightly excited by his new shoes, the little crescents of steel fitted to each cloven hoof, by the scrubbing he'd been given the day before and by the general air of business. He skipped once or twice on the stones and kicked backwards, not in malice, but out of a sense of life and fun.

They stood and admired him, coat washed and smooth, eyes deep brown and kindly, intelligent for a bull, horns short, curved and polished, his tail white and plumy. 'Now my fine fellow,' said Simon, 'are you going to win that prize?' Jasper snorted. Joady looked anxious. 'Do you reckon he might have a touch of ringworm on his nose?' he asked. 'Nay, I don't know,' said Simon, peering more closely at the wet black skin. 'But you can slap some salve on before you go.' 'Aye,' said Joady, 'flowers of turpentine and sulphur in bacon fat.' 'It's t'other way round,' said Simon. 'Aye,' replied Joady in hasty agreement as he went back into the shippon for the salve.

He led the bull out into the yard and smeared his nose well. Jasper snorted once again and his eyes rolled. 'Come on, come on my pretty,' hissed Joady soothingly as he patted the hanging flesh on the soft neck. 'Now don't lose thy ways, or get drunk, run away with a black woman or do owt daft in Skipton,' cautioned Simon as he handed Joady two shillings, 'or we'll sell thee off to Jemmy Rattle.' Shaken by this threat, Joady could only shuffle his feet in reply and smile apologetically. Martha called out from the dairy window, 'Be a good lad, now, Joady,' 'Bye-bye Jasper,' said Sally, putting her arms round his neck and whispering into his fringed ear until he shook his head, 'be good . . . win a prize, just for me. . . .'

Now Joady was on his way. He plodded slowly up a rough

track, rising steeply round the shoulder of the hill, to join the green lane over Malham Moor. His lunch was in his pocket, riddlebread folded round a slab of cheese. He had pennies for ale on the way and a bed in town. The sun was high and hot. Jasper walked at his shoulder, now and then turning his head to examine this new countryside or to bellow a challenge at some distant cattle. It was almost noon when Joady halted suddenly in his tracks, scratched his thin hair and began to laugh out loud with relief, shrill and high. It had come to him at last that Mr Stockdale was speaking in jest. He had no need to worry. There were no black women in Skipton anyway.

II

A VOYAGE DOWN ELLER BECK

'I'm putting her in for t' Show!' The taller, older lad spoke with surprising emphasis, as if fighting a battle with an unseen opponent, as if some old argument were still ringing in his head. The slightly bulging eyes, forward fall of his mouth and lower lip, and the drooping, thick-lobed ears, gave him something of the aspect of a rabbit himself. He was sweating heavily from the steep climb out of Skipton, damp and unhappy inside his thick, long-wearing black jacket, his baggy knickerbockers and coarse, grey stockings. The flattened bowler on his head, like an inverted pudding bowl, drooped above his ears.

Josh, the younger, smaller lad, peered at his blotched, pallid face under the hat. 'Take it off, Bertie,' he said, 'or I'll knock it off,' 'Nay,' Bertie reacted defensively, 'if me Mam finds out I'd get such a thunge!' He rubbed his smooth skull in apprehension. 'But th'art red-hot, Bertie . . . and just slutherin along.' 'Well you don't know my Mam. I reckon she's got spare eyes follerin' me. And she can't bide mucky clothes or owt messy. I've seen her at night-time coming upstairs backwards and polishing the steps as she went . . . That's me Mam.'

Josh whistled in response and threw his cap high in the air, leaving it to fall as a comment flat into the white dust of the road. The carelessly sewn patches on his flapping jacket were there for all to see. His mother was quite a different matter. They walked on in silence, an odd pair of friends. Bertie would be going to Ermysted's Grammar School in September. He was good at his lessons and his Dad could pay, for he was a partner in two canal boats. Josh Bargh would start work as a half-timer, probably in Dewhurst's cotton-spinning mill, doing a thirty-hour week with half days off for schooling. He was a little, humorous lad, with odd notions of mischief, 'lish as a cat,' his Gran used to say when she was still sensible, and so Josh set to work at jogging a sluggish Bertie on towards the top of the hill.

'Stop barging, Bargh!' They both groaned at the old joke. 'I'm puttin her in for t' Show, you know,' he repeated. 'What, yer Mam?' Josh made a surprised face, like a screwed-up, freckled goblin. 'No, me rabbit! It's Mop of course. . . . She's that pretty she's bound to win a prize. . . . We're all going to t'Show, me and me Mam and Dad. It'll be a right grand day wi Mop in t'marquee. . . . Are t' going thiself, Josh?' 'Nay, I don't know, I expect so. Just stop natterin on.' Josh was rather irritated by this fulsome self-congratulation and also by the doubt whether he could afford to go to the Show. 'Come on,' he shouted, tugging at Bertie's sleeve, 'to the water . . . to the beck . . . we've got till suppertime to explore. . . . Race thee to t'top.'

Bertie ran his best. Josh beat him easily and threw himself down among the sheep-cropped grasses in the narrow shade of the stone wall. Bertie sat beside him and wiped his face. Beads of sweat clung to the scanty black hairs of his early moustache. Other traffic was using this back lane from the dales, though a mere trickle in comparison with the flood now moving along the main roads in advance of the Show and nothing to the tide that would sweep over Skipton in the morning, but they had met a cart or two, a flock of sheep and a horseman.

Now they could discern in the distance the sunlit figures of a large bull and a thin man plodding along in their own cloud of white road-dust. Idly they watched from the shadow as the two drew nearer. The bull appeared to be leading the man by a slack rope that swung between his nose-ring and a loosely clenched hand. When it became clear that the man's eyes were softly closed in his round, brown face and that a powdering of white dust had not been wiped away from the eyelids for some little time at least, Josh needed no more evidence to bring on a fit of stifled chuckling. 'Why,' he exclaimed, 'yon chap's asleep.' He whistled one low note. Jasper turned his head suspiciously and snorted. Joady opened his eyes.

The hill sloped steeply before him into Skipton. He had cut across the desolate brown top of Malham Moor, seen the tarn in its hollow in the distance, the smoke rising from chimneys at Tarn House overlooking the waters, home of Mr Walter Morrison. Awake as yet, he and Jasper had wound down the edge of the limestone escarpment, past broken cliffs of white rock dotted with windblown yew trees and into the softer country below to join this road near the village of Rylstone.

Sleepy with the heat of the day, they had made slow progress down to Skipton, but at last it was visible, clear below them as Joady blinked, rubbing his eyes and wondered how he would find his way. The towers of the Castle and Parish Church stood nearest, half hidden by puffy trees, rising at the head of the town on higher ground, keeping watch over all that took place below them. The broad High Street, spine, main artery, market place and thoroughfare marched southward from Castle Gate and Churchyard to end in Caroline Square. A nucleus of streets and roads joined forces at this mis-shapen square then straggled into the distance down several valleys like a sprawl of limbs and appendages.

Pillowed against Park Hill, where Joady blinked in the sun, the staring windows of the Castle looked southwards over the town and a broad-arched gateway gaped open-mouthed into the top of High Street. The ancient road from York to Lancaster crossed High Street below the Castle, passing over Eller Beck at Mill Bridge, lined here and there with buildings, but soon rising to open country, forming fore-legs and shoulders on the plan. Both sides of High Street were hemmed in by close-packed ribs, back lanes and yards, running down to the canal and beck, where the poor of the town were crammed into narrow stone houses. All was a grid of stone about High Street, dark grey stone against the green hills, the natural, brown shade of gritstone slates, the lighter greys of limestone rubble, darkened by smoke to a uniform weight.

A newer part of town had grown up beyond the end of High Street, where the guts of Skipton lay in a curve of canal, railway and main road linking industrial Lancashire with industrial Yorkshire. There clustered mills, coal-yards, gasometers, the chimney of Dewhurst's Mill, a red-brick giant rising up two hundred feet in counterpoint to the older stone towers at the head. It excreted no smoke into the blue sky that day, for this was an annual holiday, the eve of the Show. Newmarket Street and Swadford Street stretched out West and East from Caroline Square, whilst to the South trailed the long tail of Keighley Road surrounded by the newly built houses of Newtown and Middle Town where the better off mill workers, overseers and lower middle classes laid oilcloth on the floors of stone terraced houses each with their own backyard and privy.

Joady, from his vantage point by Park Hill, had taken in the proximity of his goal and was about to put one foot forwards again when Jasper snorted suspiciously. 'What's this then?' Joady muttered, shading his eyes to look under the sunlight at the wall. Two lads climbed slowly out on to the roadway, grinning. 'Hello,' said Joady cautiously, you could never tell what mischief lads would get up to. 'Hello,' chirped the smaller, fair-haired lad in reply, 'that's a fine bull, Mister.' 'Aye,' replied Joady, 'he's a good un.' 'Takin' him to t' Show?' 'Aye.' 'Is he fierce?' Joady paused in thought. 'Not for me, nor his friends he's not.' Josh grinned, holding out a handful of hastily gathered plants to the bull. Jasper accepted them, thistles and all, munched and snorted. 'We'll see thee, then . . .,' and the lads made off sharply over the wall, to scamper across the fields. Joady scratched his head, looked at Jasper, innocent eye to innocent face, and yawned before continuing on his way.

'Look, there's t' beck,' called Bertie with relief. 'Didst' think it'd vanish?' 'Nay, it might have dried up in this heat.' Bertie took off his hat and looked sideways at the sun as he grunted with the effort of running. The grassy field sloped steeply down. Along the far side ran a row of surveyor's sighting poles and mounds of earth, tracing the line of the Yorkshire Dales Railway towards Grassington, but no one was about. The beck ran as strongly as ever, sparkling over stones, swirling in pools, overhung here and there by the cool shade of trees.

Josh began to heave off his dusty clothes, revealing bony arms with spikes for elbows, a skinny trunk and a face topped with his half-inch bristle of fair hair, shaved close agains the lice. 'I just want to lie in it and let it pour right through me,' he declaimed, splashing in swaggering nakedness to the deepest pool. 'Here's a good dubb,' he said, curling round in it like a herring, with only his white backside sticking out of the water.

Meanwhile Bertie had carefully taken off his jacket and laid his hat on top. He removed clogs and stockings, rolled up his breeches and stepped gingerly in. 'Take off thy breeches . . . clap 'em down!' 'Nay Josh, stop coddin me . . . me mother might find out.' 'Oh never mind thy mother,' answered Josh contemptuously, but Bertie sat down quietly on a stone and watched his white toes wriggling in the running stream. Josh was tactful, for once, and changed the subject.

'All right then, what'll we laik at? . . . I know – Double Duck . . . get thyself a stone.' Both boys searched for the best round smooth stone they could find. 'You go first,' said Bertie. 'Right, set yours up over there.' Josh pointed to a flat, anvil-shaped, grey boulder where Bertie carefully set his stone. Josh took aim and threw his own. It struck the anvil, missed its target and bounced sideways on to the bank. Both boys rushed to claim it, but Josh, being more careless, was there first, and so had the right to go again.

He threw, and struck Bertie's stone which fell into the beck. 'A hit, a hit! Now I've got two.' Jubilantly he danced in the water, splashing rainbows about him, holding high a stone in each hand. 'Find thyself another, Bertie,' and Bertie set up another target. Josh threw and missed. This time Bertie was determined. He charged through the shallows. One false step, another as his foot turned on a moving stone, and he was down in Josh's dubb, down and spluttering, to rise with dripping breeches and an awestruck face. Josh sat laughing at him, cross-legged in the water like an imp. 'Right Bertie,' he said, once he had recovered his breath, 'now tha can stop tittering about.' But a shocked Bertie was still staring at his soaked breeches and thinking of his mother.

Then his face lightened. 'Hey, Josh, there's summat moving.' 'Where, let me look!' 'Hush, keep still . . . there it is, cans't see?' The sun shone down, casting the shadows of two heads on to the rippling, deceptive surface of the beck. They peered closer. Something was there indeed, a grey-brown monster all of three inches long, its antennae undulating with the current, smooth, rounded carapace pressed back against a stone, the articulated legs unmoving and the pincers open. 'It's a gurt crayfish . . . come on . . . let's get it.' Four hands shot forwards, two heads collided. There was a splash as they both collapsed sideways into the water. The crayfish disappeared.

Josh scrambled up. 'It must have gone that way.' They delved across the stream bed, turning up stones and debris, thickening the clear water with released mud. Josh was searching closely, his nose barely an inch from the surface. He stopped, stared and reached down unbelievingly. 'Look here,' he said, 'look at this.' His face bore a triumphant grin. 'Look, here's a sixpence . . . that's really summat.' 'Let's see,' said Bertie, taking hold of the small coin, dull and tarnished almost black. 'Aye, it is a sixpence . . . do you think they'll take in in t' shops?' 'Course they will . . . dos't fancy a good lardy-cake?' Josh rubbed his thin stomach. 'You'd best save it,' answered Bertie automatically. 'What for?' 'Well . . . you shouldn't snaffle money away.'

Josh ignored him and splashed to the bank. 'Come on,' he yelled as he struggled to pull his breeches on over his wet skin, 'Get thy clothes up!' 'I can't go into town like this . . .' Bertie pointed at his soaking breeches. Josh scratched his head. 'I'll tell thee what, let's follow t' beck down . . . like explorers. Like we said, Doctor Livingstone! Mayhap thy breeches'll dry as we go.'

With clothes pulled on anyhow and Bertie's hat hanging precariously off the back of his head, they began their intrepid voyage down Eller Beck. The banks grew lush and green with tall nettles as they came to the beginnings of the Castle Woods, to the cool shade and stillness of their first jungle. A ravine began to open before them, a trench cutting deep into the very first foundations of Skipton. This was the moat dug by the beck throughout its long history which created in time a rocky cliff, a long, steep-sided ridge, ideal site for a castle. From this secure vantage point the Lords of the Honour of Skipton could keep watch to the North against the Scots, when they came to lay waste Craven.

This was a hidden, secret place of adventure for Josh and Bertie, thick-shaded by sycamores and maples, made treacherous by the writhing grey roots of huge columnar beeches. They fought downwards through elder scrub, struck in the face by green sprays of berries tinged with pinkish purple. The ferns rustled, dry branches snapped underfoot, then the explorers halted, whispering, only to set forward once more, slithering down mossy banks, their ears and eyes ever alert for the ruler of all hostile tribes, for Mister Barrett, the Castle agent and scourge of trespassing boys, and his band of warriors, the gamekeepers and gardeners who attended him.

'Come on,' said Josh, once the ravine levelled out a little, 'let's take a peep at t' Show.' The showfield, Storem's field, was beyond the Castle Woods, a little out of town on the far side of the ravine. A sunken pathway led up from the beckside, passing under the boundary wall through a slab-roofed gateway. The passage was now barred with an iron grille, soon decorated by a pair of sideways straining faces. They were at the very bottom of the showfield, seeing only glimpses of activity, with canvas lifting and flapping in the distance like sails at sea, the framework of poles and ropes, and the echo of shouted instructions. 'Wonder where they'll put t'rabbits,' said Bertie. 'Oh, shut up,' answered Josh, who was trying to work out if he could safely sneak in by that way next day.

Back into the deep woods, into the patchwork of leafy sunlight, the damp places scattered with curling yellow fungi, they followed the edge of the pathway. Now Eller Beck had become a long pool, canal-like and still, closed by a wooden sluice where water curled over, running down silken streams of weed into the sunshine. There was such a watery, mouldering smell that it held them for a moment, to thrust hands under the spray and soak their shirts to the shoulders.

Now the beck was properly in business as a working stream, passing from the feudal age to the industrial revolution. Dammed so as to give a good head of water for machinery, it ran in a sandy gutter beside a track and the lads splashed happily along it. Steep rocks, full of mysterious cracks and holes, rose above the goit and the withered stems of wild garlic drooped towards the water. 'Hist,' said Josh, 'look out.' High Mill, notched with forbidding masonry into the hillside, bulked in front of them, swallowing the stream.

The watching Castle rose behind it on a ridge that was now a precipice, level with the tops of the highest trees. 'They can see us,' yelled Bertie at the sight of battlemented towers and the black holes of windows, facing northwards. 'They'll not catch us . . . get into t' shadows!' Down the slope they raced. Bertie skidded disastrously into the muddy water of a seeping spring, moaned and slithered on with fresh destruction wrought on his devoted breeches. There lay safety, a level public place, the very end of the towpath at Springs Canal where a deep inky pool lapped at the very foot of the ivy-covered castle rock, which fell almost vertically from castle to canal. There, by the donkeys' graveyard, they rolled into the bushes and lay, panting for breath.

A roar like thunder, rumbling, rattling close beside them, made Josh raise his head a little. The empty iron truck tilted back up the runway. Another from the quarry, full of broken limestone, followed down the wire ropeway, tipped at a threatening angle and released its load down a broad wooden chute to strike against swinging flap-doors and tumble down again into a barge below, falling with such force that it seemed certain to smash through the inch-thick elm boards of the hold. But the barge merely stirred and settled a little, whilst a pattern of rippling waves spread across the dark surface of the water.

Josh was suddenly in a thoughtful mood. A gravestone, roughly carved with the name 'Polly' and the initials E. W. hid him from view. Although the Springs Canal had been cut, as a private venture, from the Castle quarry, to join the main

Leeds and Liverpool Canal by Swadford Street, it had not been built to full standard, for the bridges were too low for horses. They had used donkeys to pull the barges for a century but now all the donkeys were gone. Josh ran his fingers over the worn lettering.

'Bertie!' 'Yes?' 'What'll t' Grammar School be like?' 'Just like any school I should think. You know me Dad put me in for a Petyt scholarship, but I didn't get it.' 'Aye,' Josh responded thoughfully, 'but I'm not sure I'll take to t' mill. It's too slow and shut-in, worse nor school, I reckon. Not like this . . . not like holidays.' His face took on a look of desperate gloom as he hugged the gravestone tightly. 'I wouldn't mind driving a donkey . . .' 'They've got Gurner to pull the boats now.' 'Aye, that's right, no donkeys left. But they only pay a penny an hour at mill, penny an hour, that's all for half-timers. . . . And yer get no days off, no holidays, only Christmas and Show weekend.' 'You've got to earn your living somehow,' said Bertie. There was no answer to this and both fell silent. They lay, listening to the thunderous noises, until fully rested, then emerged from the bushes to saunter ostentatiously along the towpath, jackets slung over their shoulders. The towpath, raised a storey height above still water on one side and the running beck on the other, was a high perilous place, like a sentrywalk, or the narrow bridge to a dungeon keep, and here adventure was their due.

'Look out,' called Bertie from the rear. Here came Gurner. The fellow's flabby face was contorted with a horrible expression of strain and effort, eyes rolling, mouth twisted sideways, tongue hanging out. A roll of sacking, white with sweat, stained with dirt, hung over his shoulders. The belly-belt of a horse was buckled round his body, cutting into the rolls of flesh at his neck as his whole trunk sagged sideways under the labur of keeping an empty barge under way.

Josh and Bertie cringed at the edge of the path, rocking on the verge stones, as Gurner closed on them. He would pass close, very close, and Gurner might do anything. His mouth was blowing bubbles. His glassy eyes shifted to glare at them. Strange wheezing noises, like stifled curses, emerged from deep inside his chest, grunts mixed with spittle that sprayed towards them. Heaving, staring, his leather creaking at every ghastly step, Gurner reached them, almost touched them, passed them by and left behind a whiff of stale sweat mixed with dung.

Josh was as white as Bertie. 'I were getting ready to jump down into t' beck,' he whispered. 'I were near running,' Bertie agreed. But Josh clapped him on the shoulder. 'You know he's nobbut ninepence to t' bob,' he added sagely, 'but we'll keep a lookout when he's around from now on.' So they followed the towpath as it sank lower, down to the Low Mill, where Mattocks ground corn by water power. There the beck vanished under arches, to reappear in a little sunken area of ground above Mill Bridge, at the site where the first wattle huts of ancient Skipton had clustered round a ford. Deep in history now, in the days before even a Norman lord had found this perfect site for his castle, the beck had run through grassy hills that fed the sheep that gave the little settlement its name. Sheep town had become Skipton over the years.

'Let's go under,' said Josh, eyeing the rough-hewn keystone of Mill Bridge. So they bent their backs and walked their hands across the warty stones. It grew darker, for they were far below the roadway. Then the explorers looked sideways at each other. A hard, square blackness hung before them. 'Come on,' said Josh as he bent his head. The tunnel was greenish dark, cold and dank, with strange openings leading into it from above. Only when they could see a glimpse of light at the far end and hear the noise of the mill leat as it rejoined the parent beck dared they stumble forward hastily through pebble strewn water.

Heavy sunlight fell on them at last, striking on cool bodies, cold feet and blinking eyes, as they emerged into Skipton. Houses, sheds and walls ran down to the beck. Rubbish flourished along its banks, blacking bottles, scraps of paper, rusty tin canisters. They had fleeting glimpses of the back side of life, always fascinating, of carts and horses in broken-down stables, a clogger's shop with open door, a whitesmith's. Rightly was this called Back of the Becks. Then they stumbled on into the stench of a slaughterhouse, where long, slimy strands of offal trailed into the water and green flies buzzed in a stormy cloud.

Josh and Bertie soon tired of this new beck. A string of slime had wrapped itself clingingly round Bertie's ankles, whilst Josh had cut his toe on a can. When they came to a mill yard and sluice they stopped by mutual consent and sat down to put on their footgear. Bertie's breeches looked no drier, for the thick, woollen fabric was heavy with absorbed water. 'I'll have to stay away till they're dry,' he said in desperation, 'and I'll miss my tea.' 'Never mind,' said Josh. 'Cheer up! I'll buy tea!' and he proudly displayed his sixpence. 'Come back wi' me first, I'll have to see me Mam . . . perhaps.' He hauled Bertie to the roadway. As they emerged into High Street Bertie pulled his hat well down over his face and slunk along the wall in so determinedly discreet a fashion that several passers-by turned in astonishment to stare after him.

III

GRAN BARGH

'So tha's landed . . . hast'a?' Old Gran Bargh was crouched on a low stool in the gutter outside his house. In that narrow yard there was hardly room for people to pass. Josh drew back against the wall and began to sidle round, eyeing her apprehensively. You could never tell which way Gran would be. At times she was kind to him, would call him to sit beside her if they had coal for a fire and tell him stories of his father. He dimly remembered more of such pleasant times from when he was a baby in skirts. On other occasions, and this more frequently now, she would be unreasonably angry, apt to reach out suddenly and box his ears with a withered hand, shouting filthy insults at some opponent who clearly was not Josh himself.

His Gran quite frightened Josh when she sat, hunched up so on her stool. She was indeed so spiderlike, her small, round flabby body topped by the squashed ball of a head with a tight bun of hair attached behind it, the long skinny limbs, long at least in proportion to her body, and then, most compelling of all, the bright wandering eyes that stared from an expressionless face.

She had taken her stool to where just one diagonal of sunlight penetrated into that well of obscurity created by enclosing stone roofs and chimneys and touched, momentarily, the smoky wall of Roger's Yard. But that brief enlightenment had long gone and now she sat with an undazzled view of her neighbourhood, the communal water tap just beyond her, the cobbled gutter running down to the communal drain for slops, the shallow depression in the earth floor of the yard where in winter rested a permanent pool of foul water. Her vista was closed off by the yard end, all of twenty feet away, with its ash-midden and pair of privies opening wide their doors.

Gran Bargh's eyes had taken in her private landscape both inward and outward for some hours when suddenly she recognized her grandson's face passing close before her. 'Landed eh?' she said again in a surprisingly loud, strong voice for such a spidery person, 'Hast brought me mixture?' Caught spreadeagled by a bony knuckle shaking an inch from his face, Josh trembled. 'But Gran . . . you didn't ask me to.' His voice trembled, his eyes brightened in hope. 'Gran, look. What've you got there, stuck in yer apron!'

It would be charitable to talk of stripes, for they were long blended into a homogeneous dirty grey-blue by continued wear and occasional washing, but it was the cut-down remnant of a butcher's apron with a large, torn pocket stitched at the front, a kind of stomach sac, and from that pocket projected a bottle. It was a brown ribbed bottle with a carefully composed and dramatic label in angular eye-catching lettering. 'The Celebrated Anti-Cholera Mixture. Speedily allays Sickness, Soreness, Pains, Flatulency in the Stomach and acute griping pains in the Bowels. Available only from T. C. Irving of Sheep Street, 1s. 1½d. per bottle.'

'Sithee, Gran! It's there, in yer apron all the time.' Her eyes became dim and puzzled for a moment, then, looking downwards, she saw the bottle neck and, forgetting her grandson, stealthily took hold of it. Fidgeting among the jumble of objects in the pocket she brought out a small, cracked cup without a handle and poured herself a good half measure from the bottle, smacking her lips in anticipation. Her teeth were long gone, but the gums appeared to rattle. A gleam of pleasure, of one of life's rare moments, lit up her dirty face.

Josh grinned to himself and began to slip away. 'Where are

t' going?' 'To t' privy, Gran.' 'Dost want a taste . . . it'll do thee good?' She held out the cup in astounding generosity. Josh shook his head and wrinkled his nose. 'No, Gran, I don't want it.' 'It's good for thee!' She was becoming angry and insistent again. 'It'd have saved thy Dad . . . just a bit of this. . . .' She hugged the bottle to her belly, swaying a little on her stool. 'Me big lad,' she shouted, 'just a bit of this and he'd not have left me alone. . . . He died of cholera.'

'He died of drink an' a fever.' The sulky, sharp voice came at her through the open door. 'Leave t' lad alone . . . it's hard enough to make ends meet without you snattling it all away on yer daft mixtures. Wankerly old moucher!' Gran Bargh spat and shouted back. 'He were a better grafter nor thou . . . an' he died of cholera.' Florrie Bargh's only reply was a slightly hysterical laugh; so, interpreting this, correctly, as a sign of weakness, Gran shouted on. 'There's no good addlin sin, me lad died of cholera, and thou . . . Florrie . . . tha spends all thy time cronkin at back of t' wall . . .' The laugh came once more. 'I heard thou were fined for lateness at t' mill last week . . . docked tuppence at t' penny hole . . . that'll learn thee.'

No reply came to this last taunt, referring to facts which Florrie had kept secret as best she could. Now Gran sat rocking on her stool, a satisfied gleam in her eye, and sang loudly down the yard, as if to herself, a song about the mill, the overseers and the timekeeper waiting to catch latecomers. 'Behind 'er door Jimmy Patterson stands . . . slate and pencil in his hands . . . And there comes John Bonny on his grey mare . . . But I don't care . . . I'm off to Skipton fair . . .' Her high, loud voice quavered to the end of the verse and at once a couple of heads shot out of neighbouring doorways. 'Shut yer row,' they said sharply before she could begin again, and Gran Bargh subsided into her folded posture, hands still clasping her cup of mixture.

Mrs Hargreaves, whose son Lancelot worked on the barges and had been given that celebrated name by his mother in a fit of youthful imagination after she had heard part of a poem read out by a speaker at one of the Whitsuntide meetings, now wedged her impressive bulk across the yard just below Gran Bargh and let all hear her disapproval. 'She's daft, that's what she is. Why don't they take her away? She's barmy. Gives the yard a bad name!' This comment echoed up and down the walls, bringing as an answer a great roar of raucous laughter from the whitened skeleton who was emptying a bucket into the ash-midden and whose employment by Laland Shuttleworth, plasterer, was clearly evidenced on face and clothing all his working hours. His children also, squatting at his heels in the ashes and catching the dusty whiteness of the trade atan early age, as if it were an epidemic disease, laughed in sympathy with their father.

Roger's Yard already had a bad name, although only one among the dense, crawling rookery of lanes and alleys that opened like sewer entrances from between the shops of High Street, and were now, slowly, beginning to empty of their inhabitants as the town grew and improved. When the Castle would grant no building leases although manufacturing brought new hands in daily, when wages were so low that only the lowest rents could be afforded, then the one-up, one-down, back-to-back houses of the yards were overflowing with men, women and children, dogs, cats, pigs, rats and black-cocks. But times had changed and Mr Barrett was Castle agent. There were now a few empty houses in Roger's Yard, boarded up, not able to be let to a reliable tenant, whilst on Cock's Hill, rising at the bottom end of High Street, the broader streets and proper houses of Middle Town were half-built and would be finished soon. One day, perhaps, the yards would be empty, though to Gran Bargh they wre her natural home.

Josh came quietly back from the privy and went inside, signalling to Bertie to remain in hiding by the yard entrance. Gran caught the movement, rose and followed him. Florrie was banging the bars of the fire with a metal rod, attempting to shake some signs of life into the coaldust fire, half choked with accumulated ashes. The tiny square room, with its one window, doorway, slopstone under the stairs, bucket and two chairs, was hot and foul-smelling from the long, airless day of close living. Florrie sighed at the sight of her son, ran a hand through her straggling gingery hair and patted his hand absently.

'I don't think there's owt left to eat,' she said sulkily, 'we'll all end up in t' workhouse . . . and she snattles it away, old moucher. . . .' 'I'm not going to t' workhouse, never.' Gran had taken station on her stool close against the fire, back to the wall, to keep the whole room in prospect. 'They don't give red flannel for t' winter any more. . . . I'm not going there. . . .' 'Aye,' said Florrie, in surprising agreement with Gran, 'they've no idea, that Board of Guardians, an' I think nowt to them Overseers of the Poor, George Mason and the like. Do they reckon eighteen pence a week's a fair allowance for a widow to keep a child on?'

Florrie laid down the poker and stretched listlessly to touch the ceiling with her hands. 'Look,' she said, 'there's still a bite of oatcake by t' chimney. Josh can have it for his supper.' It was a sooty, jagged lump of grey oatbread. Josh eyed it and his throat heaved at the thought. 'I've already had a bit wi' Bertie Whinsby,' he said, not caring to reveal the sixpence and his plans to his mother. 'Oh, aye,' she replied vaguely, 'where hast been then?'

'We were watching t' Show fettlin up at Storems field.' 'Skipton Show,' shouted Gran suddenly from her crouched corner, making Florrie jump in her chair. 'Eighty year I've been goint to t' Show an every time I've had a basinful o' cream wi brandysnaps. I can taste it now . . . right now.' She smacked her gums. 'But you're not eighty, Gran.' Gran ignored her. 'Eighty year . . . cream in a piggie, an' big rolls o' brandysnap. My, they fair storkened in yer belly. . . . I've saved me pennies. . . .' She scrabbled furtively in her pocket.

'She's maunderin again,' said Florrie, 'she's got no brass left. . . .' 'Shut yer gob, caffler!' shouted Gran, furious at being pulled away from her dream of Show day and sweetness. At that moment the grey light coming from the doorway and bringing the earth smells of the yard to mix with the close human smells of the room, was slightly darkened by a bent figure which coughed in welcome. 'Hist Gran,' said Florrie, 'here's Bert Ridding.'

The lodger came inside, still a little bent over, in his usual way. The room was indeed low-ceilinged, but he was not a tall man, being rather square and broad-shouldered in build. As an old lead miner, who had left Grassington when the Duke of

Devonshire's mines closed down, he had become accustomed to stooping, just as he was braced for the fits of coughing that racked him from time to time. 'It's t' lead dust,' he would say, 'it's still working in me lungs.'

Bert Ridding paused, surveyed all three and nodded solemnly before saying, 'Good evening all.' Florrie pulled the second chair over before the fire for him to sit on. He shook his head and carried it off a pace to the doorway. 'Thank ye, but I'll take some air if there's owt to be got . . . a hot day.' Then a sudden grin lit his lined, sober face as he looked at Josh. 'I saw thee, lad,' he chuckled, 'mother naked in t' beck. Stark-limbed the pair of ye, like skinned rabbits. Is that thy friend in t' yard? 'Twere a hot day indeed.' Josh looked embarrassed. 'Daft lad,' said his mother in half-hearted criticism. 'Nay,' interposed Bert, 'there's no harm for a lad to go swimming on a hot day. He'll take nowt bad from it.' Bert winked secretly at Josh, one man to another.

While this had been going on Gran had poured herself another liberal helping of mixture from her bottle. Now she suddenly pushed the cup at Bert, her generosity overflowing in excitement at the prospect of brandysnaps next day. 'Take some,' she shouted. Bert accepted the cup and sniffed it suspiciously, shaking his head. 'I can smell rum,' he said, 'I've smelt it afore . . . and I'm a Rechabite, tha knows. I keep off drink . . . and thou'd best leave this muck alone.' He made to pour out the mixture on the flags. 'Nay,' yelled Gran. 'Give it to me.' And tears started to her eyes as she thrust out a sharp-jointed limb towards him. 'Nay,' said Bert good-humouredly, 'I were nobbut jesting. Though it can't be good for thee . . . and there's been no cholera about for years.' He peered into the cup with a distasteful grin. 'And there's summat here . . . what's this? It's wick. It's swimming.' He reached in a finger and lifted something out. 'Dost keep silver-fish in thy pockets?' The tapered silvery shape of the little insect, inhabitant of damp woodwork and rotting plaster, lay on his hard palm. 'It's dead,' he said, 'I reckon that stuff killed it at one mouthful. Thou'd best get rid on it!'

Gran suddenly made a snatch and recovered the cup from his hands, recoiling defensively to her stool by the wall. 'It's them privies,' she said, referring not to the yard end, though the smell from the cesspits could be felt as an undercurrent even at the top of the yard, but to the mill where she still worked as a cleaner. 'Stink gets in me throat. It makes me guts heave. . . . I boaken at it. . . . Been like that ever sin' me tonsils swole till I could swallow nowt an' that dentist cut em out for me . . . there . . . sitting by t' fire where th'art now. It stung me that bad . . . an me neck were right red. Me throat's been ad ever sin'. . . .'

'Aye . . . well,' Bert interrupted to calm her, for she was shaking as she spoke and dribbling a little down her chin, 'drink thy poison then. Mayhap it'll help.' But Gran was immune to the outside world by now. 'An all I ask is for me cream tomorrow. Why can't I have me cream? It's not much for an old woman. Show day once a year for eighty year. Little thanks I get.' Her eyes narrowed as she glared at Florrie. 'Yer slut.'

'Shut yer gob . . . you old moucher,' replied her daughter-in-law resignedly, 'an' drink yer mixture, or I'll put worms in it.' Gran spat into the fire. Bert Ridding was already ignoring the scene as he looked up sideways at a patch of bright sky above the rooftops as if he were still down a mineshaft. He took out a short, wooden pipe, stirred the tobacco in it with the blade of his penknife and sucked contentedly for a moment in anticipation of lighting it. Florrie was angrily running her fingers through her hair, pulling out nits and squashing them between her fingers before carefully inspecting the corpses and flicking them into the moribund fire.

Josh saw his chance. He slipped over to the doorway, turned, said 'I'm off, Mam,' in a casual way and ran down the yard towards the waiting Bertie. The first complaining wail of 'Josh, come here . . .' never reached him until he was surely safe out of earshot.

IV

MISCHIEF

The High Street was thronged. A traffic jam had built up at the bottom of Caroline Square, where waggons were drawn up half across the narrow corner beside the Ship Hotel and men were hard at work on a triumphal arch. Bundles of evergreen from the Castle woods were piled high beside them and a framework of rough timbers, laths and branches already spanned the roadway. Tomorrow it would be crowned with a banner bearing the proud title 'Trade and Commerce'.

The lads watched for a while, then walked stealthily up High Street, for Bertie still feared exposure, approaching Middle Row, the old Town Hall with its formal façade of shallow pilasters and applied round arches, and the cellar shops on either side, Tubber Scott with his barrels cluttering the pavement and Hurd the salt merchant. A figure was standing on the pavement just ahead, huddled deep in a long greenish-black tail coat, clicking his fingers as he talked earnestly to some acquaintance. 'Look out,' said Bertie, 'it's Tom Cherry, he'll tell me Dad of me next time he goes for a haircut. Let's go round.' 'He'll see nowt,' said Josh. He made a face and skipped past, flipping the coat tails as he went. Tom Cherry noticed nothing, being deep in his favourite topic. 'If all the folk I've shaved in my time came walkin down the High Street four abreast . . .'

But Bertie's fears had grown with every potential acquaintance or neighbour that he passed. High Street was too wide, too open, he would be running the gauntlet of countless looks, remarks and transmitted observations, not to mention the catapulted hobnails of the apprentice cloggers in Robinson's who could be counted on to fire at any promising target. He turned aside and cut back up the cobbled slope to Sheep Street. Josh had to follow him.

Now they could progress more securely and both stopped to gaze into the window of the newly-opened Lipton's grocery shop, framed in marble deeply cut with gilded curlicues, scrolls and the magic name. Behind plate-glass windows were rounds of cheese, baskets of soft, white bread, jars of shining amber honey. Josh fingered his sixpence. 'Clear off, you lads, go on!' The manager wore a complete white collar, white coat and full-length apron above his high-buttoned waistcoat. He shouted back to an assistant inside the shop. 'That order ready for the Castle yet? Biscuits, jam and cheese packed? Lord Barrett wants it right away. . . . Come on! You know what they're like.' He stalked back inside. The lads drifted on.

The end of Sheep Street was in sight and Josh quickened his pace, skipped, kicking his clogs till a spark flicked from the worn-down remnants of his irons. 'We'll go to t' Coffee Tavern,' he declared, and ran Bertie along to the very door. But there he hesitated. It was as elegant as Lipton's. The Skipton Coffee Tavern Company had been created by a group of local worthies wishing to provide respectable eating places for the ordinary townspeople who would otherwise have no alternative but to risk the temptations of one of Skipton's forty public houses. John Bonny Dewhurst of Dewhurst's Mill was a director of the company and so the Coffee Taverns were also intended to make a decent profit. But they offered a good beef dinner with all the trimmings for an honest sevenpence. The engraved glass door shone with polish and cleanliness. Inside were solid oak tables and chairs, white-aproned waitresses and an array of shining, tinned-copper urns, cylinders, pipes and nozzles for making tea, coffee or cocoa with te aid of hissing steam. Such a palace was a far cry indeed from Joe Wade's hot pea saloon, and with his sixpence creating delusions of grandeur, Josh felt an irresistible desire to live the high life for once in his existence.

'What ails thee?' A young, friendly voice called across the street to Josh. Surrounded by an array of spades, shovels, scythes and crooks, Fred Manby, the younger brother in a family ironmongery business, was bending over a part-loaded hand-cart by the side of his shop. He was a kindly, quiet man, slight and small, not yet twenty and brought early into responsibility by his father's death which left him in charge of the foundry whilst his brother Tom handled the general business. Show day had called him away from the forges. The hands were celebrating their holiday and he was needed in town.

He straightened up and pulled down the neb of his overhanging cap to shade his weak eyes from the sun. 'Is there something the matter?' he called again. Josh and Bertie had no fear of Mr Manby. They walked across to where he stood. 'We were just going into t' Coffee Tavern,' Josh explained. 'My,' remarked Fred Manby, 'come into the money, have you? You'll not be wanting the job I was going to give you, then?' 'Yes, well we have some money, at present,' Josh replied with due caution. 'But you could use some more, eh? Most folk could,' Fred Manby tugged softly at his small, silky moustache and winked at them. 'Off you go now,' he said, 'don't be afraid of Ada Birtwhistle at the Coffee Tavern, she won't bite. You can come back in half an hour, say at seven o'clock, and give my lad a hand up to Jerry Croft with this lot,' he pointed to the hand cart, 'it's needed by the Menagerie. There'll be twopence each.' 'Thanks, Mr Manby,' they replied in chorus and returned across the street.

Two girls were coming out of the Coffee Tavern. Younger than Josh and Bertie, in gingery ringlets, clean white aprons and brown buckled clogs. Belle and Annie Tawney, they went to th Catholic School up Gargrave Road, lived a mile out of town, and so were given a penny for a cup of tea in a decent place before walking home. Sister Frances had brought them all in for tidying up that morning, they had stayed to play with friends and now must hurry back to their terrace house down Keighley Road. Annie was giggling at her sister. Look,' she squeaked, opening a crumby palm, 'I've still got half a biscuit,' and she giggled again. Josh watched in disgust. If those lasses could go inside he wasn't going to be afraid. He dragged Bertie until their noses were barely an inch from the Coffee Tavern's wall.

They stood before the window, the fine scrolled lettering and the forbidding mahogany doorway. 'Come away,' said Bertie, 'they'll never let us in.' 'No! Come on,' said Josh fiercely, clutching his sixpence in a sweating hand, jacket still thrown over his shoulder. He stood his ground as the manageress advanced towards him, until her black-swathed bosom seemed to overhang his head. 'What do you lads want?' a no-nonsense sort of voice, but not unkindly. 'Something to eat, please.' The careful eyes narrowed a little. 'Don't you talk nonsense to me . . . now then,' as she surveyed them, 'you're Bertie Whinsby, aren't you? My you're in a state . . . I don't know what your Mam'll say . . .' She softened for a moment, there was a twinkle in her eye. Josh saw it and his courage rose. He gave her his most charming, impish grin.

'Want something to eat, eh? How are you going to pay?' It felt pleasantly cool now in the Coffee Tavern, and Josh relaxed his grip, holding out the sixpence. The manageress took it from him and eyed it disparagingly. 'Whatever have you been doing to it?' she said. 'Is it buried treasure, then ?' 'Aye,' replied Josh instantly and grinned again for double effect. The manageress tapped the coin on a table, looked at it again, then at the two raised faces before her.

'Right, young gentlemen,' she said, 'you come and sit down at this table here.' She led them to a half-concealed corner, close beside the serving counter, where, no doubt, she could keep an eye on them in case her kindness should prove misplaced. 'Now, what'll you be taking? Tea and cocoa's a penny a mug. Ham sandwiches twopence apiece, or we've some rice pudding left over. You can have a plate of that for twopence.' 'How about ham sandwiches?' asked Bertie thoughtfully. 'Nay,' replied Josh with scorn, 'sandwiches don't stick to your ribs . . . we'll have . . . two mugs of cocoa and two plates of rice pud . . . good uns . . . please.'

The manageress sighed, for it was still a hot evening outside and every cart passing down High Street raised a thirsty cloud of white dust. 'A right pair of lads,' she thought as she declaimed their order to the kitchen. But still, when the puddings came they were heaped high on the plates, thick with crisp burnt edges and creamy yellow skin. Josh stuffed himself and finished Bertie's for him. Then he sat back in relaxed, self-conscious leisure and sipped his cocoa as if it were vintage claret. 'My that's rich,' he breathed.

They staggered outside, sweating and happy, after thanking the manageress most profusely. She accepted their thanks with due gravity whilst bursting with laughter internally. Bertie's thoughts drained back to his breeches, for he had left a wet mark on the seat, and he heaved his hat brim low down over his face once more. They were at the heart of the town and the risk of an unlucky meeting was accordingly high. They paused by Manby's shop for a moment, to cool off and watch the completion of the double arch that spanned High Street at this, its narrowest point. It was a splendid arch, topped by a triangular frame supporting a flagpole wreathed in greenery.

'Hey, Josh,' said Bertie thoughtfully, 'what about that job for Mr Manby?' 'Aye, that's right!' Josh was all alertness. 'We're spent up now. Let's find him.' Fred Manby came out of the shop. 'Ah, it's you two,' he said, 'I reckoned I could count on you to come back for twopence. Now look,' he led them round the corner, 'here's a load of stakes. Take it up to the fairground at Jerry Croft. It's for Mrs East. She's in the big red caravan, and she holds the brass. Make sure she knows that lot's come. Then there's another load of the same. When it's done you'll get your coppers.' He rapped both heads with his knuckles. 'Ye're a pair o' mischief-makers but I know you'll do a good job.'

'That we will, Mr Manby,' Josh replied as he took the front handles, ready to steer, and set Bertie behind as the driving force. An irresistible opportunity lay before them. With pounding clogs they built up a full-blooded charge along High Street into the midst of the crowd of lads who were watching the arrival of the fair. Reckless of consequences, driving all before them amongst the train of gaudily painted waggons, they whooped and yelled with the excitement. 'Hey up! Watch out!' The shouts of anger vanished behind them as Josh swung the trolley round an inch from the hub of an enormous wheel and ran their load up the lane to Jerry Croft. 'That were Bargh and Whinsby,' said one of the lads when they recovered, 'we'll get 'em later.'

Josh and Bertie duly reported at the red caravan, delivered their load, returned with a second one, though more circumspectly, and at length collected their twopences from Fred Manby. 'Good lads,' he said, 'no trouble?' 'Nowt to speak on, Mr Manby.' 'Right. Off you go.' They ran back up the High Street to catch the last of the fun.

The end of the procession appeared on Mill Bridge. The followers were shouting to each other 'Come on . . . look at this!' A baggy-trousered fellow in a bright green velvet waistcoat came first, and then, its head nodding as if tired, with its little eyes rolling, the elephant followed, padding softly with shuffling gait. This was a creature to watch with bated breath as it lurched ponderously down the everyday surface of Skipton High Street, its tattered ears swaying, horses backing hurriedly out of its path. Ignoring Jerry Croft and the rest of the menagerie, the puny hand led his giant charge over to Kendall's Yard, an alleyway nearly as close and grimy as Roger's Yard itself, where a faded sign 'Good Stabling' hung over the narrow entrance. Somehow they squeezed her through and into the yard behind, to astonish the wives sitting at their doorways with the sight of a wrinkled, grey flank suddenly darkening their vision at a foot's distance. Several lads followed her down the yard to the clutter of privies, pigsties, and stables beyond the houses. There her green guide backed her round and into a stable where, with melancholy head scraping the beams above the doorway, she ran her trunk stealthily out through the half-door like an escaping snake. Now one other smell was added to the collection of farmyard odours that habitually drifted along Kendall's Yard and seeped out at intervals into High Street to the surprise of passers-by.

A real gang of lads now, they all ran back together to watch the fun in Jerry Croft, to stand in the dusty grass by the wall and enjoy the spectacle, the free show put on before them, as canvas and tents were pitched, beasts fed and watered and odd show people dressed in an engaging mixture of richness, colour and dirt, good enough for lamplight, ran everywhere in organized confusion. 'Harry Stevens is here,' said Bertie uneasily, with memories of the time when he had first been at school and Harry's gang had persecuted him for weeks and at last had tied him by his jacket buttons to a door latch, kicked the door and run away. They had chosen the door of a crotchety old fellow, often plagued by boys, and Bertie's ears still throbbed at the memory of the revenge against collective injuries wreaked on his individual skull.

'Hey,' yelled Harry, seeing them and waving a gorilla arm. 'Come on. We're laikin. Come on . . . black plum. Last one in's a dirty dish-clout.' At the back corner of Jerry Croft stood an old post, worn and wormy, broken off at a good sitting height. 'Bertie Whinsby can be first plum,' said Harry Stevens, with his nose for the frightened smell of a natural victim. His yellow face took on the mockery of a smile and all the lads laughed, throwing themselves about. 'No,' said Josh indignantly, 'it's not fair. Ye're always on to Bertie. I'll be plum.' He took his place defiantly on the post. 'Let me be champion!' Bertie's pride was returning. 'I'll show 'em,' he added, and began to take off his belt. 'Have you got any string, Josh?' With a length of string he tied his breeches clumsily around his waist and then, for greater safety, ran the cord around his shirt buttons.

'Right, come on!' shouted Josh suddenly. A swarm of lads advanced on them, swinging belts, flailing their arms with switches of grass and nettle in each hand, aiming to hurt the plum and his champion, to beat them black and blue, yet not be trapped themselves and forced to the same perilous station. Josh was not one to sit tamely and be whipped. He covered his face with one hand while with the other he clubbed at his attackers with a branch of elder. Bertie fought white-faced and determined, with a courage he owed his friend, to capture a fresh plum and sit him on the sacrificial post. Josh grabbed Harry Stevens' ankles. Bertie dived at his face. Over they all toppled. The fighting grew more abandoned. Shouts and yells resounded from a tangled knot of struggling boys.

Suddenly, without warning, a missile landed right among them, quelling the noise, splattering on arms and legs, stinging faces. The heap of twisted bodies stilled. Muck slid from startled faces and plopped to the grass. A bearded burly face frowned down on them. The riding boots, the billy-cock hat, the dung-bucket still swinging in his knuckled hands, showed he was a man in authority here. 'Clear off . . . get out . . . little buggers,' he croaked, shaking a fist at them. The lads began to disentangle themselves and drew back. All were silent until there was a noise of tearing and a sudden gasp from Bertie as the boy who was kneeling on the edge of his breeches rolled away, pulling at the fabric until down they came, taking shirt buttons with them. 'Oh no,' shrilled Bertie. 'Shut up . . . get going,' yelled the man and stepped forward, flourishing his bucket. Clutching torn, dung-spattered, stained breeches to his stomach Bertie ran with the others and wept as he ran at the thought of his mother.

The gang had scattered. Josh and Bertie stood, gathering breath and composure, at the lane entrance to Jerry Croft, by the side of the Red Lion Inn. The shadows were lengthening with the evening. They watched an urgently self-important woman approaching with a black bag and whiff of carbolic soap. It was Mrs Speakman, the midwife and manufacturer of the celebrated Skipton relish. 'You know what she's got in that bag?' whispered Josh. 'Yes,' replied Bertie. 'No, you don't . . . I'll tell you. That's how she carries babies about. . . . If there's someone as wants rid of a baby, and someone as wants one . . .' He winked at Bertie. They both chuckled. Mrs Speakman eyed them suspiciously, then stopped. 'Is that really Bertie Whinsby?' she asked loudly. Bertie hung his head. 'I don't know what thy Mam'll say. Think on it. It's nowt to laugh about.' Bertie hung his head lower. Mrs Speakman marched on.

Behind the Red Lion lay its yard, typical of the old Skipton inns, which had long been farmsteads, with their own pasture land rented from the Castle and still needed stabling and grazing for all the beasts that were brought in to market. So

Jerry Croft was a temporary grazing ground for the inn when not rented to the fair. Bertie led a retreat into the yard. Each stall and shed was full, the heads of horses and cattle projecting at amazing angles from every corner. Heaps of new hay, sacks of oats and buckets of meal were tipped against the walls wherever space could be found. Close by the gate into the lane stood an old, crazy, open pen, hardly more than a token of enclosure, and in it stood a bull.

Josh looked at the bull. The bull ignored Josh. 'Hey Bertie,' he whispered, trying to divert his friend, 'that's the one we saw this afternoon.' Bertie nodded gloomily. A fly buzzed Jasper's head and the bull lashed his tail in irritation. Josh was idly contemplating that tail, for it was a fine creamy-white with a long plume to it. A thought came to him. Here was something to cheer Bertie. His eyes twinkled. 'Look at that tail,' he said, 'he's a fine un . . . bet he'll dance to music.' He whispered for a moment into Bertie's ear and a slow grin of enlightenment spread over his friend's face. 'Aye, he'll dance all right,' he agreed.

Clang! A bell note hard against the ear would make anyone jump. Billy Hull, the bellman, laughed at their surprise and rang another single chime directly in their faces. 'Sowin gape seed, eh? . . . Been slowin away in Jerry Croft, I'll be bound.' He waved the bell handle at them, one finger on the clapper. 'Bertie Whinsby! Thy Mam told me to look out for thee.' He eyed Bertie up and down. 'An' she'll give thee what for. Hast seen thiself? Hast a seeming-glass at home?' He grinned broadly at the disconsolate Bertie, then took his ear. 'Come on . . . best get it over.' 'All right,' groaned Bertie, his shoulders drooping in despair. 'So long,' shouted Josh and, winking heartily, 'don't forget . . . right early in t' morning . . . tha knows why!'

V

THE HONOUR OF SKIPTON

A hand was scratching with fine steel nib across a white page that lay in the ring of lamplight on the desk top. Within the narrow enclosure of Conduit Court, buried deep in the thick-walled masonry of Skipton Castle, light was cut off early and so the oil-lamp stood burning on a square mahogany partner's desk with its array of drawers and store of precious documents. The writing was crisp and authoritative with bold downstrokes, for the hand belonged to Robert Bell Barrett.

Castle agent, cousin to Lord Hothfield, who owned the Castle and all its lands, Mr Barrett was uncrowned king of Skipton, for the young lord visited the town perhaps twice a year, and then generally only for the shooting. Mr Barrett was practical master of all the castle lands, of wide farms, grouse moors, woods, and all attendant and appurtenant rights, master of over half the town of Skipton and landlord to shopkeepers, craftsmen, householders. If the Honour of Skipton, as its lordship was anciently known, had outward and visible form, it was in Lord Barrett, the by-name his subjects gave him, and his hooked nose, not in the fleeting passage of its titular owner.

The Barrett family attended church every Sunday in an open carriage, were met at the door by the Reverend Cooke, M.A., in his gaiters, bow legs and glossily shaven cheeks, whereupon Mr Barrett would raise his grey silk topper half an inch and proceed inside to meet his God and his social inferiors. He read the lesson as if he were giving instructions to a rabble of beaters half across a grouse moor.

Mr Barrett's estate office was at the very heart of the old Castle, its windows looking out on to an ancient yew tree, growing strong in its octagonal stone box in the centre of Conduit Court as it had done for over two hundred years and making the castle's core a dank, cool, mossy place even on the sultriest of days. On the wall behind his desk a protrait of old Lord Hothfield, in his ceremonial dress as Mayor of Appleby, silk breeches and all, looked down on him approvingly. Mr Barrett thought of tomorrow's Show in Storems Field, which he had rented once again to the Craven Agricultural Society. His eyes wandered to the great estate map hanging on the end wall of the office, its varnished surface yellowed with age, the dim light barely revealing where the faded pink wash of some skilful draughtsman's brush had proudly outlined all the Castle lands. Mr Barrett looked at his watch and turned back to the paper before him. He did not like speechmaking, but since Lord Hothfield, who was, of course, patron of the Show together with the Duke of Devonshire, would not be attending, it fell to himself to speak on behalf of the Castle and the Honour of Skipton at the formal luncheon next day.

An apologetic shuffling of papers in the back corner of the room distracted Mr Barrett's attention once more. 'Mr Giles,' he called to his temporary clerk. 'Yes, Mr Barrett?' 'Haven't you completed that lease business yet?' 'Well, you see Mr Barrett, I'm not yet accustomed to the work. I can't quite see what the lawyers want to do about the new office, and . . .' 'That's enough. I've told you before. I won't have long explanations.' 'But, Mr Barrett!' 'Now get on with it, Giles. I told Fell that lease would be ready by today. We put the rent up to sixty-five pounds for lead-works, shops and all. The office is covered once it's built.' 'Yes, sir.' Giles made to retire to the standing desk again, brushing down the curly black quiff over his forehead with the heel of his hand as he went.

'Oh, Giles!' 'Yes, sir,' again. 'We shall be meeting here in a few minutes. Get out the Show catalogues and papers, will you, please. Witham should be up soon with some biscuits and wine. Let him in. Then you'd best clear out.' 'Thank you, sir.' 'I didn't say, "go home", you idiot! There's that lease to finish.' 'Sorry, sir.' 'You'd best take it into the closet. There should be a candle. Don't go till it's finished. I want to sign it tonight.' 'Yes, sir. Shall I come in when I've got it all done?' 'Of course.' Mr Barrett turned back to his speech. Giles began to open drawers and extract their contents. He placed the Show catalogue, its cover displaying a remarkably mixed collection of engraved beasts all staring woodenly at the bottom corner of the page, on top of the pile and laid it, tactfully, six inches from Mr Barrett's elbow.

There was a kick at the door and Giles hurried to open it. Joe Witham stepped inside, a tray full of bottles, glasses and covered plates in his hands. 'They sent this over from t' house,' he said loudly. The Tudor wing of the Castle, where the Barretts lived, was known as the house. 'Put it down, Witham,' said Mr Barrett brusquely. Witham looked around the room, saw no alternative, and dropped the tray noisily on to the desk top. 'Now it's landed, I'll be off,' he remarked. Giles straightened the tray's disturbed contents, saw Witham out, collected his papers, and retired to the closet.

There he sighed long and loud, ran his fingers through his hair until all the black waves became curls and made a half-peevish, half humorous face at the dreaded lease. Giles sat down in the mouldering, winged armchair that for some reason provided the only seating in this store-room and prepared himself for a final effort. He had never meant to take this job. He still hoped for far better things after such a good education. Really, he ought to go to college, or be articled to a learned profession, an architect say. That would be something. He liked buildings. This Castle now, he'd read all about it in Dawson's *History of Skipton.* He could close his eyes and see it growing, from the first Norman walls with their arched entrance, to the finished medieval fortress of mighty round towers linked together in the shape of a letter D, the courtyard in the centre. In the yew tree's branches a guardian owl perched nightly, hooting softly from this vantage point to all the Castle ghosts before flitting noiselessly away over battlements and turrets, stone-slated roofs and silent empty rooms to its hunting in the woods below.

The Castle must be full of ghosts. Giles opened his eyes and examined the shadowy room with caution. All those dreadful deeds in the history called out for ghostly vengance, from the murders of black-faced John Clifford to the poor wretches once locked in the airless, lightless, heatless dungeon forty feet below. Through a slit window Giles could just perceive the far side of the courtyard, the entrance to the new Tudor kitchens and the carved arms of the Clifford family, with two stone supporters, fish-bodied, winged, dragon-headed, a pair of wyverns. They swam before his eyes. He was tired.

A distant thump and crash echoed from the canal basin below the Castle as the very last load of rock was dropped into its barge that night. It was a familiar sound. Giles heard it again through his dreams as he drifted in time and the thunderous noise took on a deeper significance.

The past of the Castle is not only built all around him, but he is part of it, his imagination stirred by the rumble of distant guns. He can hear the sound echoing towards him from Cock Hill, which closes the southern end of High Street. His mind is back in the days of the Civil War, when the old yew tree was a new sapling, and the Castle came into its own once more as a King's garrison, held through three years of siege. General Lambert, from Calton Hall near Malham, has seized Craven for the Parliament and set up a battery of guns that now pound the castle walls. Henry, Lord Clifford, fifth and last Earl of Cumberland and thirteenth Lord of Skipton, is dead in York of a burning fever and the castle is held for King Charles by Sir John Mallory with a garrison of three hundred men. Lambert's guns are sounding in rolling volleys from Cock Hill, the balls whistling the length of High Street to strike with a crash and splintering of stone on the old, round towers. Soldiers of the Royalist garrison fill the Castle, crowded into this dense sculpture of stone, seemingly almost solid from the enemy's side, with its sloping battered base leading smoothly into the curves of clustered towers broken only by the black arrow slits. Skipton Castle is an engine of purpose, and now at war once again as Sir John Mallory guards its honour and from the forward watch-tower his sentries look down over the low stone houses of Skipton town.

Sir John himself, his officers and ladies, take more comfortable refuge in the Tudor wing, where deep bay-windows survey only green lawns sheltered by the bailey wall. Sir John's ladies rustle up the splendid staircase with its heavy, black-oak, turned balustrading and wander freely along the great gallery.

On and off the siege has been in progress, part blockade, occasional cannonade, since 1642, and life in the Castle has not been much discommoded. The Royal Standard flaunts high over Skipton town and troops mingle with the local folk on market days. Then comes the hour of reckoning in the North. At the battle of Marston Moor Prince Rupert is defeated by Cromwell's horse and driven from the strong city of York. Although the battle is lost and won in three short hours, the fighting is not over, and darkness falls on the King's Yorkshire infantry locked in bloody conflict, hand to hand with Cromwell's men, refusing to surrender, until Lord Fairfax, his face streaming with blood, forces his way between them, knocking up swords, hoarsely yelling, 'Spare your countrymen.'

Royalist stragglers flee widely over the northern countryside, bringing the sad tidings to remote garrisons and with them the threat of final defeat and capture. Early in the spring of 1645 General Lambert returns to the siege of Skipton, which is to be pressed home to the hilt. In the castle Sir John is stirred to action. A surprise attack is planned, a sally over the snow to the rebel camp. Major Hughes, the second-in-command, saddles up at dawn and leads his troops, grey steel helmets wet with morning damp, cloaks pulled close around their shoulders, out of the gate and across the ford by Castle Mill to sweep secretly round the town and attack by surprise.

Mid-morning, snow-covered huts and tents, the smoke of camp-fires, lie before them as they canter, yelling, sword in hand, at the enemy. It is soon over, the surprise is complete, bringing a reward of a hundred prisoners, sixty horses and a load of plunder. Jubilant and light-hearted, they trot back along the high road to Skipton. But General Lambert has learnt serious warfare since he left the quiet dales. He comes after them in hot pursuit. The Royalists are too encumbered to escape and it will be a fight, yet another of the many thousand local skirmishes of the civil war.

The prisoners are loose, attacking from the rear, seizing weapons. Major Hughes's men are too few. They break and run. Twenty dead lie at the roadside. Now the chase thunders into Skipton, horses lathered with the gallop, booty flung desperately into ditches. They roar into the end of High Street to see the gates of safety opening in the distance, the Royal Standard floating above. Farmers scatter. Errant cows flee irritably as the hurricane sweeps by. Horses whinny at the fun. But the arched gateway looms closer, the drawbridge is down over the dry moat and the sound of their hooves changes to a hollow echo running over the wooden bridge and under the sheltering stone. Then they are out into the mid-day sunlight of the defended bailey and able to draw breath. Major Hughes falls from his horse, heavy with wounds, to be carried into the Tudor house and die in the night.

Lambert's men are battering at the door. Guns from the flat leaden roofs of the Castle fire down at them and they must retreat, for the broad High Street is an open murder way, exposed and merciless in this rain of iron. Lambert will settle this business now. It has plagued him too long. He brings up fresh batteries, entrenching them on Park Hill and begins a last serious bombardment coupled with a close blockade.

Windows smash in the Tudor house. It was not built for war. Sir John withdraws into the old Castle itself. His rooms look out on to Conduit Court, the stripling yew tree and the water cistern fed by a conduit from outside the walls. At many times of the day he will visit the sentries stationed on the roof of the watch-tower and gaze out towards his enemy below in an attempt to prejudge his plans. Assaults follow and are beaten off, but there can be no help from outside.

It is hot weather for armoured soldiers in August, when the garrison remember their peaceful lives in the hayfields of the past. They swelter in idleness until, suddenly, a great assault is on them. Two large parties of Parliament men have crept close to the walls, slipping between the huddled houses of

Skipton town. Guns are dragged up High Street to face the gates and hurriedly defended against castle fire with barrels full of earth. Smoke rises from musket and cannon. The local householders, who have come rather to look on this siege as an entertaining spectacle, cower in their cellars.

The Parliament men come on, teeth set, pikes beside muskets, ladders ready for the walls. They are on top. Sir John's men break and run for the Castle. Sir John rallies them to counter-attack, but already another party of enemies are in sight beyond the gatehouse. This fight is lost. The garrison have no refuge but the old Castle, and there they are close shut up.

Their enemies now hold the church and the Castle outworks, walls and bailey, and also they have cut the conduit. For water the garrison can rely only on rain caught by their roofs, and there are many wounded among them, salt food to eat, and crowded quarters. Yet still they hold on, whilst every other fortress in Yorkshire falls and they are left alone.

The end comes in the bitter month before Christmas when food and water are desperately short and there is no hope, no prospect of relief. Sir John agrees to treat for terms whilst the old courtesies still hold good. His courage is recognized, his desperation acknowledged, in a written agreement made and kept. This, after all, is a war between neighbours. So, a siege of three years now over, Sir John and his men march out of their long-held Castle 'according to the honour of a soldier, with Colours flying, Trumpets sounding, Drums beating, Matches lighted on both ends, and Bullets in their mouths,' as the signed treaty provides. Proudly, but sadly, they march out of the old gatehouse and down the long stretch of the entrenched High Street, vanishing round the corner and into the hills in search of what Royal forces remain in arms. The Parliamentary officers lead a search party through the now deserted Castle, and as they dig out the last bottles of wine from the cellar their laughter echoes along the passageway into Conduit Court and round the stone walls that enclose it.

VI

AN URGENT DELIVERY

The page was turned, the lamp trimmed, and Mr Barrett worked at his speech undisturbed for ten minutes. A heavy sound of footsteps on the stairway made him raise his head. The booming kindliness of the voice revealed who was his first visitor, for the staircase and landing created a kind of stone echo-chamber outside the estate office. 'Do not trouble yourself, my friend, I know my way,' said the voice, obviously speaking to the gatekeeper, and went on in response to some inaudible comment. 'Indeed? The rheumatism still a problem? Here's something for it.' Mr Barrett made a wry grimace as he laid down his pen. He rose to his feet when Mr Walter Morrison, Justice of the Peace, and formerly Member of Parliament for the Skipton Division, burst impetuously into the room.

'Ah, Morrison, you're early I think.' 'Yes, yes, it was such a glorious evening up in Malham I decided to walk. Fifteen miles. Marvellous. The air's like warm red wine and I did it in the four hours just.' 'What, again! You're a glutton for exercise. Do sit down.' 'Thank you. I do like to celebrate my health, to feel it working, you know. After that touch of T.B. It's gone now, thank goodness.' Only a slight cloud had passed over Walter Morrison's beaming face at this reminder of past weakness, just as it did when people introduced him as 'former' Member of Parliament. His massive bulk, air of confidence and jovial enthusiasm, like a giant schoolboy, clad in a flapping suit of fine grey herringbone, seemed to shrink Mr Barrett a little by contrast, to make him draw into himself even though he stood, rigid as a poker, in his own office.

Walter Morrison had barely time to take a glass of wine when a knock at the door was followed by the immediate entry of a round, rosy face fringed in grey hair. 'Ah Walter, Walter, I nearly caught you. But you were off too fast for me. Like a pudding after the roast beef. . . . Well, well, Barrett, how many brace was it?' The Reverend L. B. Morris shook hands with both of them, stroked his plump bowler with plump fingers, laughed, and accepted a glass of port wine.

'Where were you shooting?' asked Walter Morrison. 'Over on Rombald's Moor.' The Reverend Morris pointed eastwards towards the hills that isolated Skipton from the rest of Yorkshire. 'Barrett slaughtered them by the hundred, by the hundred!'

'How many more d'you expect at this time of night? There seems very little to discuss.' Walter Morrison had tired of this subject and was keen to get down to business. 'Only Dewhurst promised to come. I don't expect more. The secretary gave his excuses. We'd better wait for Dewhurst.' 'He's generally on time,' said the Reverend Morris. 'He's been a pestilential nuisance at meetings since he gave up managing his mill,' retorted Mr Barrett. 'I read your Primrose League speech last week,' Morris commented to Walter Morrison as they sipped their wine and waited. 'Why, when Skipton's misrepresented by that damn Liberal, speechmaking is all I can do.'

'Damn good thing this Boer business is almost over,' Barrett interjected. 'Threatened to split the country.' 'All due to Kitchener. He's the man,' said Walter Morrison, bursting into speech. 'I discovered him, you know, and pushed him on as best I could. I had to sit on the War Office's heads for months before they'd see. But it worked at last. Now the Boers will have to accept peace.' 'The Liberals are putting one or two nasty stories about.' 'The beggars won't work, but they will whine. What do people expect in war? Our men were dying like flies, yet we took the Boer women and children, concentrated them in camps and gave them the best food and clothing we could. Do you think I would be party to any cruelty?' One look at Walter Morrison's face as he heaved indignantly in his chair was enough to answer such a question for all eternity.

Slow steps and the tap of a cane could be heard on the stone landing. 'Well, well,' said the Reverend Morris, 'here's Dewhurst come,' and he went to open the door. John Bonny Dewhurst edged sardonically through the entrance and took an upright chair by the window. 'Who's chairman?' he asked in a dry, old man's voice. 'Barrett, I suppose! Good evening to ye.' 'Have a glass of wine, Dewhurst,' replied Mr Barrett. 'What's on the agenda? Morrison here'll be up to one of his wild schemes I suppose. That's it, eh?'

Walter Morrison was already rumbling ominously in his chair. 'You didn't think the Yorkshire Dales Railway a wild scheme. You've taken shares enough in that.' 'Aye, and there are calls enough on 'em!' 'You can't make an omelette. . . .' 'I know, without breaking eggs. But see you don't break mine.' Mr Barrett cut into this exchange of pleasantries. 'If you'll bear with us, Dewhurst, we have the revised show programme to consider.' 'Oh, I'll bear with ye, I'll bear with ye.' Mr Barrett, somewhat put out, rubbed his hooked nose with two fingers. Dewhurst's small eyes twinkled in cold amusement and his goat beard jutted forward obstinately. The man who had built up the biggest mill in Skipton was not one to let even a single point slip by him unacknowledged.

'Morrison wants us to give time for a new item. He's buying a ploughing machine and he wants all the farmers to see it at work.' Here Walter Morrison, alive again with enthusiasm, lurched forward and interrupted him. 'When we investigate agricultural production we find a vast gulf, an enormous difference, between the good and the bad farmers. Now education is needed in such a case. Education is the answer. But you can't take a dales farmer who's been on the land, man and boy, for half a century and stick him in a schoolroom. You must teach by example. This is what I aim to do with my farming. If an agricultural show is not the place to . . . to make this public, then nowhere is! Now what I have ordered from Leeds is a new, experimental engine to pull the plough.

It may not succeed, though I am confident that it will, and I can afford it. But it will certainly demonstrate what can be done, and at the Show every farmer will see it. They must be given the lesson.'

'How does this do better than the present machinery?' asked the Reverend Morris who had a glebe farm himself. 'We've had steam ploughs and traction engines for years and years.' 'Ah,' answered Walter Morrison, relishing the chance to explain. 'This is something different. You'll see that it is lighter, and therefore more stable, and also more powerful. It is powered by a new paraffin-oil engine. It should confound all the sceptics utterly.' 'A horseless plough for a horseless carriage,' remarked Dewhurst into his beard. 'Why yes, yes! That's it, Dewhurst. You wait and see. I'm having it painted red and green! If we bring a traction engine or two in from the railway workings, perhaps a steam plough, we could show the world a magnificent display of farming machinery.'

'Then what the devil are we sitting here for?' John Bonny Dewhurst spoke sharply, as he was able, poking with his stick as if to explode the image of progress that swelled visibly before Walter Morrison's eyes. 'Give it five minutes' time after the prizes. That's enough. It can blow up then.' 'It won't blow up. This merits far more than five minutes at the end, I can tell you. It could bring a revolution in our Craven agriculture.'

As Chairman, Mr Barrett put the question. 'It's almost nine o'clock. We must reach some conclusion. Are we going to stretch the programme, or not? Any display should go before the prize-giving so that people stay to watch it.' Walter Morrison nodded vehemently in agreement. 'But the machine is not yet arrived? Is it sure to be here tomorrow?' 'They are bringing it from Leeds by canal. Bateman should be looking out for it, if he's sober,' answered Walter Morrison. 'Perhaps we should send to find out?' 'Yes, that's a sensible proposition. I'll send Giles off right away. It should only take him twenty minutes to the canal yard and back.'

Mr Barrett went into the closet, shutting the door behind him. Soon the remainder of the committee could hear his voice raised in anger and a soft, defensive reply. The voices emerged on to the landing. 'Now get down to the canal office right away! Send up Bateman or whoever's there. Don't go to sleep on the way or I'll sack you this minute!' 'Yes sir.' The young voice was swallowed by the staircase. Mr Barrett returned.

'Trouble with the hand, eh Barrett?' remarked Dewhurst slyly. 'I'm afraid he'll have to go. He's a dreamer. Just now I found him asleep in an armchair. Supposed to be working on a lease.' 'Perhaps he was tired,' suggested the Reverend Morris. 'Tired! Got no excuse to be tired! We'd best go over the rest of the programme till Bateman comes.'

They had almost finished the business when a voice shouted thickly just outside the door, 'Where's t' lantern, show a light!' heralding their visitor. 'Please Mr Bateman,' Giles's voice implored, 'stand up and come this way.' A door handle rattled. 'It's not in there,' persisted Giles. 'That's a store-room. Come this way, please.' Barrett threw open the door and old Bateman walked blinking into the light, almost full into his arms. 'Go home, Giles!' said Barrett sternly, and Giles fled.

Bateman gazed round the room as if to challenge its existence. He was a stiffly built, broad, short-legged fellow, over seventy, with a much-dented brown bowler still on his head. 'Take your hat off, man,' said Barrett. Walter Morrison laughed. 'He's been drinking again. They don't call him "cold water Albert" like his son.' 'Drunk,' retorted Bateman, 'I bain't drunk. Had a warmer, nowt more!' He put his thumbs in his waistcoat pockets, shuffled his feet astride into a pub orator's stance and asked, 'Now then, what do you grand gentlemen want wi' me at this time o' night?'

'Mr Morrison has a cargo coming from Leeds,' said Barrett. 'Aye.' 'It's an engine from Fowler's, needed for the Show, and should have been here today.' 'Aye!' 'Is it come?' 'Nay. And it'll not be here tonight. There's a fly-boat bringing it. They've orders not to run after dark.' 'Why not?' 'Well, seems clear enough to me – they might run into summat.' 'No . . . you damn blockhead. Why has it not come?' Barrett spelled out the sentence coldly. Old Bateman at once turned sullen at the tone. 'How should I know? It's delayed. That's all.'

'Now, listen to me, Bateman,' said Barrett in his most

lordly manner. 'I won't take insolence. Give me a straight answer or I'll be on to the Company Agent first thing tomorrow morning.' Bateman shuffled for a moment and grumbled into his beard, then he answered. 'Sorry, sir, I were just a little warm. We put a good man on to run that order through. He's very reliable. You can be sure for certain, he'll get her in on time. Should be in t' yard first thing, but I'll go right back and check.' 'You're sure about it?' 'Can't beat this man, sir. He's a regular worker. Best we've got in t' whole barrel load.' 'All right, Bateman. Clear out now. But send word as soon as it arrives.' 'Aye, sir.' Old Bateman worked his way on to the landing and half-walked, half-fell, down the dark staircase, thinking to himself 'I'll bloody murder that Badger if he don't get back first thing. Why the hell did I take him on?' His thoughts would have been blacker still if he could have seen the present condition of his good reliable man.

VII

BADGER

Only a last remnant of sunset light penetrated the boat cabin through half-curtained windows and the glass bull's-eye let into the roof. There the shadowy, slumbering bulk of Badger snored and mumbled away his afternoon's drinking. The fire had died to ashes. His clogs mouldered in condensed dampness on the step, where he had kicked them off, one behind the other like leaders in a procession of giant black beetles invading the boat. All was still on the small fly-boat, lightly built for speed, where a half-sheeted iron tractor, in red and green livery, projected from the hold, its chimney gleaming in burnished metal.

Some warning, some sense of urgency, stirred that heap of flesh in the cabin. He twitched, rolled over, and opened one black eye. Grey, obscured, flowing greasily like the canal, his mind slithered over muddy thoughts. Disconnected images floated behind his eyes, gigantic waves of shuttering coal roaring towards him like breakers on the sea, the skull of a horse with staring eye-sockets, a battered pewter snuff-box, a woman's slow smile. For a moment he slept again, the visions absorbed back into the turgid stream. Anxiety, a touch of consciousness, stirred him to waken once again. Time, work, order, these imperatives would not let him lie and for a confused moment he felt the stone walls enclosing him, heard the harsh bell, the tramp of warders' feet and clang of iron doors. Badger shook in a spasm of fear. He rolled and fell heavily to all fours on the floor. A fragment of tufted rug rose up before his face.

'Bloody hell,' he groaned, the drink still swilling in his hanging belly, 'where's that lousy, coddin' lad?' The clouded anxieties suddenly crystallized into a sharp fall of certainty. Had not gaffer Bateman ordered him to be back in Skipton this night? He had been held up on turn-around at Leeds, waiting for the arrival of the tractor and then loading it; a legitimate defence, though it would never prevent the gaffer from giving out a good tongue-lashing. But then there had been Bingley five-rise, with it being so dry, and to cap it all he'd lost his ticket for settling the dues with the lock-keeper. So Badger had gone off for a drink, leaving Perse the miserable boat-lad to search for the lost ticket, with orders not to go to bed until he found it, 'or I'll larrup thee till thy skin flakes off,' Badger had added from the towpath.

There was a tap at the window. Badger heaved himself painfully upright and bent to glare out, wiping a smudged circle of condensed breath away with the heel of a grubby hand. The lock-keeper confronted him, obviously taking care not to lose the opportunity of collecting his ticket for ever. 'Bugger you,' intoned Badger as he went to the ticket hole by the cabin door, but no scrap of pasteboard had re-appeared like magic inside that dusty wooden slot. The lad hadn't found it, then. Badger vaguely recalled cursing him as he returned and throwing a gobbet of dried mud at him where he sat, shivering, by his sleep-hole in the bows, too scared to go to bed, too tired to seek the ticket any longer, a thin, hopeless wraith in the sunset.

Badger rubbed the dusty spikes of his black hair and then felt all over his crumpled body as if to verify its continued existence. Then he stuck a stubby finger inside his waistcoat pockets, felt something and rolled his little black eyeballs in a peculiar giddy motion at the discovery. There it was, the lost ticket. He opened the cabin door, coughed arrogantly at the misty air and flung the card at the lock-keeper's feet. 'Thanks,' said the man, bending to gather it in.

'Damn you,' replied Badger under his breath. He drew the

knotted warning whip out of his belt and looked greedily at the departing back, then swung round and cracked it down fiercely on the cabin roof, at the same time running over a miscellaneous bag of insults under his breath. . . . 'Skin yer head, yarmin runt. . . .' His malice shifted as he thought of the pale lad, still shivering on his pile of sacks and old ropes at the fore end. 'Get yer backside movin,' he yelled out loud, cupping one hand to carry his voice the boat length, and cracked the whip once more. The lock-keeper was already out of sight beside the misty locks.

A dirty, white, staring-eyed face emerged from the fore-cabin, nodding with lack of sleep like an old tortoise head. Perse lived most of his life at the fore-end, crouching knees to chin, yawning, but impelled by fear to be always ready at the crack of a whip, to leap off and tie up, to untether the horse and harness it, to run ahead to lock gates, to cook dinner, clean up a little, shovel coal feebly, and occasionally even take a spell at the tiller. 'Hey you! Pig head!' Badger's summons drew him to scramble over the angular cargo, his mouth open. Then Badger smiled with his little flesh-embedded, rosy lips and nodded with his head to the interior of the main cabin. 'What's t' eat? What's for dinner? What's cooking?' A pause, then, 'There's half a pail dinner left,' the boy whispered sullenly. 'Speak up, speak out, don't squeak. Dids't say "pail dinner"? Not again, not wi' ham bones, saw bones and sawdust. I'll larrup thee . . . get in there. Get inside. Hide.'

Badger's clog scraped menacingly on the foot board as the boy scuttled past into the cabin. Like the rest of the boat this had sadly declined since the days of its glory when it was last overhauled and painted up as the home for a honeymoon couple who had begun married life running cargoes, but soon moved on to better things. The miniature cooking range had rarely been cleaned in Badger's time and flue and chimney were rimed with a thick coating of soot, generating a good smoky fug when the doors were clamped shut and Badger had flung himself down on his bunk. Above the crock cupboard one last plate hung in its wire frame, the edging of lacy china broken off in jagged lumps, a distressed survivor of the original shining row. The painted door of the cupboard, which let down as a table, bore a scarred picture, an architectural nightmare of a castle in red and brown, rich in impossible turrets, battlements of all shapes and sizes, slit windows, pointed roofs, all set in a sylvan green landscape of enamelled colours. The painting had been commissioned by the young lovers. The clog marks and obscuring grime were Badger's.

The boy set the blackened kettle and tin-lined pail with its wire handle and bamboo grip to bubble quietly on the range, poked the fire in a disheartened fashion, sat down on the side bed to watch it, and, lulled by the noise, the slight movement of the boat and the absence of Badger's eye, for only his corduroy legs were visible, like the shaggy tree trunks darkening the doorway, fell heavily asleep. Badger himself was dreaming at the tiller. No amount of hurry would get them to Skipton much faster. He would beat the horse, he wouldn't mind beating the horse, but it required an effort and Badger was happy in his dreams.

He was thinking of Izzie, of Isabel, and of her body, the fat white thighs, the soft rolls of flesh at her waist that he delighted to pinch and make her squeal, the smooth skin inside her arms, short firm neck and bulging breasts. Badger's hand tightened on the carved octagonal stem of the tiller and the whole assembly, black-tarred rudder, rudder post covered in geometric circles of yellow and red, the tattered remnant of a grey horse-hair tail fixed to the ram's head, all shook as if the boat itself were relishing his dream as a dog twitches in his sleep.

A splattering of the eddy shook Badger back into consciousness. 'What the hell. . . .' Here he stood at the tiller and they weren't moving. Late already, miles away, piles away. . . . The damn horse wasn't harnessed. Chewing its guts away behind the hedge. Where was that lousy belly-aching lad? . . . Badger lurched into the cabin and whipped the loitering boy up on to the towpath.

The horse itself was not much of a creature, an indeterminate breed, cob or vanner, low built to get under bridges, thick-boned and heavy muscled, a greyish subdued brown in colour, damp with condensed cloud. The broad back and rump were almost bare flesh, scarred and black with the sores of a lifetime. Badger looked it in the eye like a cold executioner as he checked the harness. The horse knew Badger. It kept still, avoiding wrath, not even shaking its head in reminder that it had not been fed.

Badger checked the harness perfunctorily, neck collar with short brass hames projecting, broad martingale running down the chest, between the legs, and then the traces themselves. These hung either side of the horse's flanks and carried a row of painted egg-shaped bobbins to prevent them catching. The traces were fixed to a curved swingletree, one at each side, and from a hook at the centre of the swingletree the snubbing rope ran direct to the T-shaped fore-end stud in the bows. Badger untied the snubber and cracked his whip as he stepped aboard.

Now the half sun seemed almost level with the canal surface, sending streaks of glowing red jumping along the ripples of the boat's wake. The motto of the Leeds and Liverpool Canal Company was 'de ortu ad occasum', from the rising to the setting. Its boatmen were guaranteed such sunsets whenever the clouds lifted.

Badger was musing angrily as he cracked the whip and th horse took the strain. He'd have to keep going till after dar well into the night. Old Bateman was bound to be fratc either way, but if he arrived too late next morning he mig well lose this job and not get another. He leant heavily sideways against the triangular end of the cabin, his belly resting on the painted cratch board. He had fixed up to take Izzie to the Show. He had to be there ready and waiting. Izzie expected it. The tiller quivered, vibrating in the water, created tiny ripples which flowed away behind the boat, mingling with those mighty parallel lines carved by the whole flat-bottomed passage of long elm boards through the broad canal.

The thought of Izzie made Badger frown at his post as the horse plodded indifferently on along the towpath and Keighley drew nearer, Tolkein's Mills, Waterloo Mills and Elliot Street Mills. It brought to mind his rival, his opposite; irritating, quick Lancelot! The very recollection of the name brought a sneer to Badger's cherry lips, his instinctive response to the thought of Lot Hargreaves, though Lot was no match for Badger in physique and indeed would have been swallowed up by Izzie. But that dangerous image, creating a jealous ache in Badger's loins, had to be hastily wiped away. For Izzie liked a good show and Lot's boat was a sight for all to see, a glory of primary colours set against newly grained and combed light oak planking.

To see Lot Hargreaves in the early morning was to watch a gawky figure skipping along gunwales, over cabin roof, paint-pot in hand, the smell of sweet linseed oil about him, a chalked snap cord in his hand for better definition of his line painting. 'Snap,' he would twang the cord, a miniature cloud of chalk dust would rise in the air and Lot would set to on his new line with eager zest. The floorboards in his cabin were fresh with new red paint. His water barrel was thick with roses, even the underside of his tin pail glowed green and red. Coiled ropes decorated his decking, whitened daily, and an intertwined swan's neck of plaited rope swayed between rudder post and rudder. This image of Lot the artist, ever touching up the extension pieces to his chimney, the brass-rimmed cans that kept smoke from his cabin, faded as Badger gave a deep convulsive chuckle, for Lot was a defeated rival. There was something in Lot that he could never stomach, that worked in his belly, making his muscles tighten just to see that twisting face.

The stream of Badger's dreams flowed before him along the dark canal, leading him under bridges, swirling, bubbling, viscous as ale, raising behind his eyes a dark past and lustful future, blotting out the humdrum present. As the canal neared Skipton it bent round under the darker shadow of the hills, tailings off from Rombald's Moor. Badger was dozing peacefully at the tiller. It was almost midnight. The beer-and-pail dinner in his belly had settled down to a steady, brewing fermentation. He barely noted the bridge name-plates through his half-closed eyes as the horse plodded under Milking Hill Bridge, crawling to the right soon afterwards and very soon again edging left under Humblethorpe Bridge, before sweeping on towards Snaygill Stone Bridge at a steady pace. Leisurely, relaxed, Badger fingered the tiller like a woman's neck, rested his broad chin on his spare hand, and eased his mind into its flow of images.

The first shock tumbled him forward. A crack, a thud as the whole boat shook. His chin caught hard on the cant-board. His neck jerked with the full weight of his body before he toppled down, slithering on to the coal-box. The boat swung out across the canal. As the snubbing line jerked, tautened, and grew slack again, the horse stepped sideways and stopped in surprise, fearing some new act of malice, whilst the swingletree thrashed sideways and back, its point digging into the towpath, the harness twisted across its rump. 'What the hell,' cursed Badger as he rolled over on to his belly, 'we've bloody hit summat . . . must 'ave . . . hit summat . . . I'll murder the bank rangers, murder 'em. Hell, bloody hell! . . . If it's damn holed I'll leather 'em. Where's that bloody brat? . . .'

The fore-end was drifting out still, and with a way on. The boat rebounded against the far bank and Badger, almost on his feet, his head resounding with pain, was thrown tumbling over into the cabin once again. He crawled out on hands and knees along the cabin, over the cargo, to reach the boy and the rope. 'Get hold,' he shouted, without time even for the customary blow or curse. 'Heave in! Heave . . . Heave!' The lad laid an ineffectual hand on the snubbing line. Badger was putting all his weight against the rope, urgently striving to get the boat straight and under way again. The bows began to move, slowly, but in the right direction, until the horse – feeling uneasy at being used as the fulcrum for this leverage of thirty tons and Badger, and without clear instructions or training for such an emergency – began to back cautiously down the towpath, so slackening the rope once more. Dragging the whip from his belt, Badger cracked it furiously over the gunwale. 'Get on there . . . hey up,' he shouted. The horse hesitated. The whip cracked once more and the boat was under way again. Badger felt his neck creaking with every movement of the boat. Waves of pain swept up into his head. 'Bloody Hell,' he groaned. There was no sense in trying to get back now. He'd have to tie up and look at the damage in the morning.

VIII

THE RED LORD

The select committee meeting in Skipton Castle had broken up hours before Badger's accident, and as Barrett was seeing his visitors to their carriages, the rearguard of circus performers and sideshow entertainers, were still trailing into the High Street. They became dark shapes against a darkening scene whilst William Haigh, the lamplighter, went his slow rounds and added a heavy penumbra with every scatter of pulsing light.

One of these shapes, a young, slim, tired man, was followed by a shuffling creature on a chain, like a large dog. He came into the lamplight and paused for a moment's rest, the creature revealing itself as a half-grown, brown bear cub that sat down abruptly on its haunches as soon as the tension on the chain was released. 'Ah, Jacco,' said the man softly to the bear, 'dove, dove andiamo?'

He was clearly lost. Josh emerged from the shadows where he had been watching events since Bertie retired. He had no desire to go home. 'Will he get much bigger, Mister?' 'Eh?' The bear-keeper was slightly taken aback. 'The bear? Jacco? Yes, yes he will be as big as I. . . . Tell me, young man, I seek a place to sleep, do you know where is Navvy Charlie's house?' 'The navvies' doss-house?' Course I do. It's in Union Square. Come on, I'll show you! What's your name, Mister?' 'My name is Gianpiero,' replied the quiet young man, giving a tug at the lead, 'but here I am called Johnny, Johnny Itie.' 'I'll show you to Union Square,' Josh confirmed, 'but I'll bet they're full at Navvy Charlie's.' Gianpiero sighed, shrugged his shoulders and followed Josh into the labyrinth of black yards that ran down from High Street to the canal. In such narrow spaces full night had existed for an hour, seeming to be a permanent, and necessary, state of affairs, a recognition that here were lives and circumstances to be hidden.

Union Square, as poor and black as any yard, was entered through a tunnel smelling of middens and old cabbage leaves. Hens scavenged over the cobbles. An air of seedy decay pervaded the whole gloomy space, imprisoned by the three-storey rubble houses with their heavy slab roofs. A hundred years before, when first built, it had been a hopeful place, the focus of a textile worker's community, with hand-loom shops in the attic of every house, lit by the many small panes of the wide windows. Now it crouched in the shadow of Dewhurst's Mill, the hand-loom weavers no longer existed, and its inhabitants scraped such a living as they could.

Josh led Gianpiero across the slippery cobbles towards Navvy Charlie's doss-house. Just as they came into the dim light of a gas-lamp he drew back to circle warily round a shapeless heap of a man bending to drink at the square's water tap. It was Gurner again, and he was mouthing to himself, dribbling the water, mixed with spittle, down his chin and twisted neck. 'Two today . . . reet today. Two boats a bob, a bob, a bob . . .' and he danced a little ghastly jig around the tap as he celebrated his day's earnings from boat haulage before shuffling off into the darkness, grinning to himself as he went.

A whole gang of rough lodgers were sprawled on the earth outside Navvy Charlie's door. Those were good times for lodging-house keepers, what with the Yorkshire Dales Railway being built, and the Show in prospect, in addition to the regular customers, canal men, travellers, weavers out of Lancashire looking for work and cowmen or shepherds in town for the market. Gianpiero, Jacco and Josh approached the gatekeepers with due caution, but the navvies proved to be in a kindly mood. There were joking remarks of course, but

Bacon Billy, Hector, Frowsty Sam, Little Charlie and Tony Corrigan each had a pound or two of wages from the railway building in his pockets and were about to go up town for a drink. All was rosy benevolence. The mood might well be different when they returned. Josh left Gianpiero in their hands, remarked, 'I'll see thee tomorrow,' and slipped away.

Navvy Charlie too, squat and bald, was in welcoming mood. The money flowing to his pockets had brought almost a glimmer of humanity with it. Gianpiero was promised a share of a mattress in return for payment in advance. Jacco proved more of a problem since no one cared to share a blanket with him, but at length, and for a further consideration, he was accommodated in the cellar with the hens, though securely chained to prevent him taking any liberties with his unwilling hosts. That done, it only remained for Gianpiero to find his fellow showmen and claim a pitch near Jerry Croft. The party of navvies were beginning their amble up-town and took him under their wing. 'Come wi' us, lad. We'll take thee to t' Red Lion.'

The journey was relatively uneventful. Along Swadford Street they admired the floral chamber-pots in the windows of Baldisaro Porri's china shop and even perched for a moment on the broad window sills to sing a verse or two of raucous serenade. Mr Porri himself soon opened the window of his house above the premises, poked out his sleek grey head, like a tortoise from its shell, and admonished them. 'Will you please go away and sing somewhere else, or I will pour something over you.' Good humour still reigned. The navvies laughed at his courage, nudging Gianpiero with their elbows, joking, 'He's one o' thine. Say summat. Go on!' But Gianpiero bit his lips. He knew the social divide between prosperous shopkeeper and itinerant bear-keeper even if they were once fellow-countrymen. So they passed on, until at last they were at the portals of the Red Lion, its regular façade and new sash-windows belying the old farmhouse inn that existed behind the door. The navvies directed Gianpiero down the lane to Jerry Croft towards his negotiations with the fair boss for a good pitch. All the oil-lamps in caravans and tents were glowing, giving the fair encampment the aspect of a crowded village that had slipped in wholesale from the wilds and was not yet settled into regular pattern.

Many people had gone to bed early that night, though a refreshing coolness had settled from the hills after the warm, sleepy evening, and the moon was rising over the rooftops. But in the sandy-floored bars of Skipton's multitude of public houses, the Hole in the Wall, the Cock and Bottle, the Brick Hall, Fountain, Wheatsheaf, Old George, Black Horse, the Castle and twice as many more, including of course the Red Lion where Joady Thacker now lodged, the night ran on in a glow of gaslight, a hubbub of conversation and the outbursts of laughter that punctuated such talking.

Joady slipped quietly into the bar after seeing to Jasper's food and drink, with the relatively firm intention of having only one mug of ale before bedtime. A noisy group of navvies had just barged in and were buying drinks. Joady's eyes watered in the fog of tobacco smoke. He brushed at them with his sleeve and signalled cautiously to the barman. A tin tray piled with greyish-white titbits, swimming in oily juice, rested close to his elbow. He viewed with respect the chunks of cooked and salted tripe. A scratched paper notice beside them read 'Customers are invited to help themselves.' Joady reached out a pair of fingers, pincer-wise, but the touch, the spongy resilience, slithering against his skin, made him withdraw in distaste, though normally he was partial to a plate of tripe and vinegar.

Someone took his elbow, a burly, sunburnt, out-of-doors kind of fellow in leather gaiters. 'Look who's here!' The voice was thick with ale, resonant and rough-edged. 'Here's Joady Thacker. He's a quiet 'un, I'll lay to it.' Joady opened his mouth. 'Shut yer gob, Joady,' said his friend rapidly. 'Come on, take a pint,' and he thrust a mug of ale into Joady's hand. Joady took it and raised his eyebrows. 'He's a quiet 'un,' Tom Waller repeated. 'Now, if I can get him to laugh.' A roar of ribaldry, doubt and amusement greeted this. Joady blushed, as all the attention was focused on his face, and lowered his eyes to his mug. 'If I can get him laughin' . . . then ye'll all buy me a pint . . . agreed?' 'Nay, he's grinning already,' the protest was instantaneous. So was Tom Waller's reply: 'Aye, but he's that quiet, I tell thee; he won't rarely laugh . . . and then he waits till he's wi' t' beasts into t' shippon.' The guffaws that followed were taken as a sign of contractual approval by Tom and his circle alike.

'Well,' said Tom, 'you know I'm looking out for another place?' Whistles of amazement followed. 'Th'art not thinking of leavin Mr Brockall, Tom?' Tom winked, and laid a finger to his nose, grinning inwardly all the while as he leant, splay-legged for ease, against a settle end. 'Whatever for, lad? He's a good maister, ain' he?' 'Aye, well enough,' Tom replied loudly, 'but you listen to this . . . Not two months agone we had a pig die of sickness, and he salted it down and gave it us to eat.' His listeners shook their head disapprovingly at this meanness. They were waiting for the joke to begin. 'Then a few weeks after, we had a calf die, out in t' close, and we had that calf for supper, roast, baked and stewed over a week or more.' 'Come on, Tom', said a fat-bellied acquaintance, 'get on wi' t' tale. No more o' Mr Brockall . . .' But Tom pursed his lips, paused for effect, frowned gloomily, and then continued as he had begun. 'We'd been given dead mutton an' all.' A further chorus of disapproving whistles echoed round him. Their interest was roused, they would wait a while for the joke. 'And then,' said Tom, pausing, as all eyes rested on his face, 'and then, but two days past, when we'd nobbut finished the calf . . . his wife died . . .'

Silence, as they all looked in amazement at each other. Tom took his time, looking round at their pop-eyed faces before he added, 'so I reckoned I'd better leave him.' Then slowly the chuckles began developing, growing at length into a roar of laughter. 'Tom lad, tha were pulling our legs . . . come on! Must have a pint of ale for that!' But 'Wait a minute,' said the bellied acquaintance, 'what about t' quiet lad . . . what's he doing?' Joady felt the eyes on him again. His face had crinkled deeply as an old apple once he knew that Tom wasn't serious about this tale. A very low, coughing chuckle came from his lips. The foggy tears were in his eyes. Tom slapped him heartily on the back. 'Well done, pal,' he said, 'drinks for thee an' all, I reckon. It takes summat to make thee speak i' public!'

'That's an old tale. I've heard it many a time.' The voice that spoke from the shadows had a different tone to it, in manner of speech, slight edge of accent, even in its rhythm it was an educated voice. It was not a voice of the gentry, not southern educated, round and flowing, but perhaps came from an old grammar-school boy, a schoolmaster perhaps or a Methodist preacher. Though what such a one might be doing in the Red Lion bar raised a host of questions.

The voice spoke again from the shadows, with a slurred edge, as if from drink. 'That tale of dead beasts is older than my hat . . . and that's a thing that's seen its day! Dead beasts indeed!'

The voice, nearer now as its owner advanced into the gaslight, was sarcastic in its higher knowledge. 'Tales,' it said, 'I know tales to make you all tremble and retire to your beds to dream nightmares. It is black night outside. A time for fear to walk the hidden earth. A time to bolt doors, bar windows and shut out . . . whatever is there. I have a tale of night and terror to recount, if some kind cowman or drover will fill this pot. . . .' A bony hand with half the forefinger missing held up a quart pot, empty, for only two drops and a lazy trickle of froth fell into darkness from the circle of lamplight. The face appeared, bony also, pointed like an arrow toward the bent hook of the nose. The thinness was unhealthy, perhaps a result of hunger, but the eyes were bright and grey, so perhaps it was illness alone that made the man so talkative, so quick and so sarcastic.

'Right,' answered Tom Waller, and warmth filled the bar once more, a conviction that there was a world of beasts and men outside, not some walking terror of the night. 'Right, have one o' mine, and earn some more. It's right good addlin!' The skeleton man nodded fiercely, seized the mug and drank it dry almost without pause. 'I have many tales,' he said, and his eyes half closed as he spoke. 'But all of them are strange, frightening, eerie, best not told before a night's walk home across the dark. Will you hear one . . . ?' 'Aye,' came the reply in unison as they gathered around him. 'Get on wi' it, lad,' added the belly, 'As you wish. Put another quart in front of me . . . I guarantee to chill your bones.'

'Cast your minds back, my friends, I conjure you, to those dark and gloomy times when the Norman invaders of our land filled its green valleys with slaughtered men, ravished women, the blood of children and the smoke of burning farmsteads.' He paused dramatically to allow time for a few additional stragglers to gather into the crowded group around him. Clearly he loved to have an audience, whatever he was. 'The greedy hands of those fierce tyrants tore apart the old loyalties to fix a new and heavy set of shackles on the people. You know of them here.' He pointed a thin, bent finger through the hazy room to the dark window, to where the Castle loomed over the town as its first lord, Robert de Romville, had done eight hundred years ago.

'The legend I tell of those days I heard in another dale to the north of here. If it is current coin hereabouts you can inform me, and purchase a quart of ale in compensation. Either way I will drink it.' He sighed as he drained his mug, pushing it languidly into the nearest hand for refilling. 'This lord was a fierce tyrant indeed, a bloodthirsty ruffian, wretch and cruel oppressor of his subjects. Many of those dalesfolk had died in the harrying of the North, no family but grieved for some loss, yet still he drove them with his sword. He was a tall, thin, bitter man with a red beard and a rounded belly. We know him only as the Red Lord.'

The thin man's glassy, fevered eyes swivelled round to inspect the attentive features of his audience. 'Believe the story if you like,' he said, 'it is a good one. I sell the best words in Craven, and for no pay but ale.' A chorus of guffaws echoed round the room, but he raised his hand for silence. 'Such a man has fear at his heart, fear of the evil he has knowingly committed, to mingle with his greed for power, for domination, for the sight of fear in others. Thus he drove them to build for him a stronghold, his castle, eyrie, keep and dungeon. Leering across the quiet land he had them build it, close against a glaring white cliff, on a craggy, twisted spur of rock. With bleeding hands and broken backs, driven by the lash, his serfs hauled the mighty logs to build this wooden keep, a dungeon to watch over them.

'At length it was complete, standing like a vile, thickened

arrow to threaten heaven and earth, refuge of fear, place of domination, staked around with pointed beams, densely roofed in close-packed, shaggy heather, watching through slitted openings over the scattered crofts of the village folk below. At times the Red Lord would summon them closer, to witness his "justice", as he called it, and they would gather at the command, a shuffling crowd in their ragged homespun, following the broken pathway that wound upwards beside the rock edge until the Red Lord appeared above them. Like a monster from an old legend, fiery bearded, grinning in savagery, he would survey his flock, fierce delight in his eye, before pronouncing doom.

'The folk lived thus for many years, subject to the care of such a shepherd, whilst he scraped in his taxes, dispensed his justice, exacted every scruple of his rights, and relished the execution of his punishments. So grim, distant, unfeeling, he seemed, that the villagers lost all belief in his humanity, preferring rather to dream that his ugly body had become the seat of an even uglier evil-spirit, that he was a devil-lord, immune against the Cross itself.

'Of pleasures, save the delight in pain, he had but few. The chase he loved, whether of wild red deer or tame men who sought to flee his lands, or perhaps the village girls when desire took him. Alone he lived without kin, master of his band, the mailed robber-soldiers who shared his keep, his household watched over only by one old man, a slave from his ancestral homelands, speaking with a raven's croak, black-cloaked, taciturn, sideways-looking and shuffling hamstrung about his duties.

'The Red Lord cared for no man living. Alone among created beings he showed some kindliness to his hound, Fenris. Perhaps he welcomed the kindred spirit, the equality of savage temper, vicious nature and cruel delight which seethed inside this brindled wolfhound bitch, waist-high to a tall man, yellow-fanged, slit eyes glaring like a wolf herself.

'To this hound he gave one great task which she undertook with fierce intelligence. She must guard, protect and cherish the herd of bristled swine that were the Red Lord's pride. No other swine were like them, barrel-bodied, bristle-backed, horn-snouted, striped and spotted. Villagers said they were the devil's children as they invaded the crofts, rooting up precious worts and beans, the children's winter food, sowing terror where they came, devouring the fruits of a season's hard labour, bringing starvation and death. Yet if some creature showed defiance, if some poor crofter or shepherd lad so much as raised a hand to throw a stone, then was Fenris mightily angry. With lips curling, jaws gaping and a growl that shook the earth she told the quaking folk that the Red Lord's swine were fiend-guarded, hell-protected.

'The Red Lord rode out to hunt with the Fenris hound by his side. Smoking mists swirled clear from his windswept keep. Wild birds screamed high above the cliffs as he rode towards the high moorland in search of prey. Hot sun beat down on the red-faced lord as he quartered the hills, but no game could he find. Deep anger welled inside him. At length he came at a walking pace into a sheltered valley dotted with grazing sheep. The peaceful scene jarred his restless spirit. Feverishly he set loose the great hound. "Get them . . . cursed grass eaters, down them, destroy." Onward sprang the hound of hell, biting, driving, snarling, howling.

'Out from the shade of an overhanging rock sprang a slender figure, fair hair swirling in the sunlight, barefoot with a crook, a shepherd girl. The Red Lord's eyes grew narrow as his hound's. He spurred his horse towards this unknown girl who was frantically running to gather her sheep. She heard the hoofbeats and, turning her head, saw with terror the grinning mask of the Red Lord as he bore down on her. Springing desperately away she ran, hair streaming, across the open valley, the dread rider close behind her. His leather skinned arm swung out to grasp her, whip in hand. Almost he touched her flying hair, almost he had her, when, at the last moment, she swerved and vanished down a narrow gully, full of broken rocks. The Red Lord reined his horse back on its haunches in his fury. Shrilly he whistled to his hound, pointing out the trail of the fleeing girl, setting the hunter in pursuit of fresh, young game. "Go Fenris! Go! Bring it down!" he yelled.

'The hound leaped forward. The Red Lord followed,

stumbling over the rocks, his anger still growing. Then he came upon an angled cliff-face and laughed out loud at the sight before him. Trapped, her fair face reddened with the chase, head pressed back against the rock, the girl was screaming at the snarling creature which threatened her, at the bared, yellow fangs and the long claws that tore her clothing, leaving red gashes on the white flesh beneath. The Red Lord stood, watching, and laughed once more, a sinister blaring sound. He licked his lips in relish. The hound jerked upwards, spurred on by his presence, tearing again at the girl's clothing. She screamed once more, and as she screamed, flailed with the cross-headed crook at Fenris's snarling face. A sudden howl of pain followed. The hound drew back, jerking, and as it did so, the sharp point that had pierced its eye, digging through the eyelid like a fish hook, tore at the eyeball, wrenching it from its socket. Blood spurted on the stones. The hound crawled away, howling dismally.

'Never before had rage so swollen the Red Lord's heart, not even amidst the heat of battle when sharp swords stung his own flesh. He advanced towards the shrinking girl. She cowered with arms before her face, attempting no resistance as he drew nearer. Blood from the injured hound mingled with the red of her own torn thighs. The Red Lord's hate sought some instrument. At his feet rolled a jagged star-pointed rock. He stooped. It rose in his hands without effort. He raised it high above his head, then paused to savour the look of terror on the girl's agonized face. But a strange, disturbing acceptance lay there. She smiled. He brought it crashing down. Through the thin hands, held up as if in prayer, it drove, cracking bone, tearing flesh, battering into the quiet face. Another blow followed, and another, more rocks thudding into the defenceless heap of flesh now lying by the cliff. His harsh, panting breath echoed around him as he rested at length from his labours. The Red Lord built a cairn above her broken body, a monument to his murder.'

The thin man paused for breath. 'Very good,' said Tom Waller, 'that were a right horrid tale. I can see there be maisters and maisters.' A laugh followed, but 'Hush', said someone, 'clap up, Tom. He bain't finished.' The thin man looked slightly pained. 'No, my friends,' he said, 'the story's only half done. I beg you not to spoil it before it's over.' A chorus of shushes ran round the room as he resumed his pose, eyes to the ceiling, and opened his mouth.

'The girl was well known and loved in the village, a child of old parents, related to most families by blood or marriage. She was missed that night. Rumours spread buzzing through the beehive huts. The villagers took note of the Red Lord's bloodstained hands as he rode home that night and of the wound his hound had suffered. Sullen looks followed, as suspicion grew to certainty, but these he expected and ignored. Soon sullen anger changed to defiance, to tight lips and clenched fists, held before him in the manner of a people driven too far.

'Within two days the Red Lord heard tidings that his prize boar was missing from its sty by the castle gate. His anger had been strangely simmering ever since that bloodstained day of hunting, not calmed by the kill, and nerves like raw knives jerked through his body at the recollection. Then came a trembling serf to report a loss which could not be hidden from him. The serf died, but what of the guard, what of his most faithful follower, the great hound who had never failed him. Summoned by his call, Fenris ran bounding to greet him, only to be met, strangely, unexpectedly, with curses, blows of the whip, kicks from heavy-mailed boots, and finally a volley of stones and staves. The dog cowered in a corner of the yard, like any whipped cur, but red-fire of anger, equal almost to the Red Lord's own, flared in her eye. She growled softly in her throat. The Red Lord answered with a blow. A howl of pain, more piercing than ever before, dinned in the Red Lord's ear. Fenris leapt back with a tremendous shaking of her great muzzle. A sharp splinter from the billet of rough wood that had struck her now hung from the wounded eye socket. Blood began to flow down her coarse pelt. With a second howl the dog slipped away, out of the keep, leaping the barred gateway, her cries echoing across the open hillside as she vanished. She was not seen again that day.

'Through the long, interminable blackness of that dismal night the sound of howling, now near, now distant, kept the

Red Lord tossing in troubled sleep. At dawn the hound was found beside the gate. She would not venture inside, but snarled in fury at any who dared approach her. The Red Lord himself came to her. As he approached, the hair on her shoulders rose in evil recognition. The Red Lord snapped his dry fingers. "Come here, carrion," he called softly. The hound snarled in answer. "Come in, then, you bitch. Damn you." Once more she snarled, as the Red Lord broke into his accustomed curses. But even this charm failed to work, for she was not one of his serfs, to be compelled always against her will.

'At length he sent for meat, a slab of raw flesh hacked from some peasant's tax-sheep, and flung it at her feet. The dog sat like a spectre before him, empty eye-socket glaring horridly, like a great red blister. He took one step forward. With a howl of fury the hound snatched up the meat and fled away from him to the hills.

'That day the Red Lord hunted alone and in a thicket came upon his prize boar, lying still amidst buzzing flies, an arrow through its neck. Red mist of anger filled his brain at this further injury, until he seemed to float in a sea of blood. His furious gallop home, loose-reined, across broken hill slopes, was matter for legend, and even as he reined in his horse he shouted vicious orders. Pitiless-eyed, squat, the foreign guards marched into the village bearing a summons that all men, all with merest strength to draw a bow, must be before the castle gates at dawn next day.

'In cold, grey light the menfolk gathered, from strongest boys of twelve to hale and hearty old men, herded into a defiant, muttering circle by the guards, whilst behind them fluttered a wider circle of tearful, trembling womenfolk, their babies and little children. A bitter, unfriendly wind blew from the cliffside. The first edge of sunlight cast the black shadow of the Red Lord's keep heavy across his waiting people. At last the gates opened.

'He stood before them, cursing, a grey axe in his hand. "My boar is dead," he said, "dead by bowshot. Either you give me the guilty one, or else," and he looked maliciously round the assembled crowd of women as if choosing his threat with care, then flung out a mailed hand to point at a small fat boy, naked, pissing in the dust at his mother's feet, "every male child shall die." A wail rose from the women, a low moan of anticipated grief. Some of the men clenched fists, drawing breath, but they were unarmed and guarded. "Well!" said the Red Lord harshly, "have you no tongues? Bring the murderer to this place by dawn tomorrow – or feel my vengeance."

'He was about to turn away when a figure, old, bent, hooded, stepped forward and made a sign as if to speak. "Who's this?" rapped the Red Lord impatiently. "Do you know something of this? Speak, old man, before I send you to a timely death." Slowly the old figure raised his watering eyes to the reddened face before him. "Hark to me, Lord," said the old man, "for these eyes have seen many things. They have seen my free folk made slaves, hunted like deer, slaughtered before their homes. They have seen your evil doings these heavy years." The Red Lord laughed and swung his axe carelessly, hissing close above the old man's head. "Silence! Stop these whinings. Tell your tale or die." "Aye," said the old man, feebly shouting into the wind. I'll tell . . . for they say those about to die are sometimes blessed with second sight." His quavering voice grew stronger. "And I say to you, Lord, that if you spill one drop of infant blood, if you so much as make one child cry in fear, then, as sure as Christ is the hope of little children, your strength is gone from you forever." His voice trailed away in mumblings. He made no move to flee, but stood, face raised to heaven, his tale told. The Red Lord raised his axe. Then the old man turned absently to go his way. With one swift kick the Red Lord felled him to the earth. Down fell the grey axe across the old man's back. The Red Lord stood back to watch. At his feet the hooded prophet floundered, hissing for air, then died into silence.

'Howling continued throughout that second night, a fearful howling from the hills and a heavy, oppressive silence from the village below. The Red Lord dreamed of corpses rising from the battlefield. Then came a compelling vision as he sweated, cold among his furs, a distant view of his own keep, high, in a staring white sky, raised above an awesome cliff which seemed to go down for ever into screaming blackness.

All was still. But round the tower circled nine black ravens, and as they circled they cawed nine times.

'This vision came to him many times during the night. He woke, sweating still, with the memory of it graven in his mind, though why it should bring such fear he could not tell. As his old servant shuffled in with morning ale he called him to his side, driven by a compulsion to speak of what he had seen, and retold the dream. The servant drew breath slowly. "Beware, master, beware," he croaked, "all is not well. I see the hand of Odin and the old ones. I fear the gates of cold hell beckoning and the last fire blazing over hall and keep. Go not forth this day." "What words are these?" growled the Red Lord, rising in his fury. "More whinings." He struck the servant to the rush-strewn floor, then stood above him, belabouring him cruelly with a whip as he rolled to and fro in the mire, hands shielding his face. At length the Red Lord desisted, tired from his labour and with other toil before him. Dawn was in the air as he strode to the gate.

'The timbers creaked. Heavy bolts were drawn back, and the Red Lord departed from his secure fortress, clad in mail, helmeted, sword in hand. His horse was led towards him. He swung easily into the saddle. A stern smile was on his face. He would deal with that rabble of serfs for ever. Nothing within him now gave warning of the doom he would face. No subtle instinct whispered to him of how despair and hate had driven his old servant to rebel so that now he limped bitterly through the keep, flaring torch in hand, the fresh blood still streaming from his whip-cut head.

'The gates were wide open, the waiting serfs gathered in shadow below, shivering in the dawn cold, and with more than cold. Horse and rider appeared, giant-like against a clear sky. They halted. A hiss of breath alone broke the silence. The armour-clad body swayed sideways in unconcealed astonishment. The bloodshot eyes glared down at rows of fat corpses lying in the dew. Nine to a row, then nine, then nine. All his swine were slain, all of them, and laid out in ritual before his own gate.

'A great roar resounded from the rocks as he swung his horse downhill, charging into the swaying crowd below. He curled from his saddle snatching a babe from its mother's arms. The high thin wail of fear that rose about him sank deep to his heart as if it were his own. At that very moment, a low sound of amazement spread through the scattered serfs about him. Faces turned towards the light, hands pointed. The Red Lord jerked backwards in his saddle, unable to resist the impulse.

'He saw a great pillar of smoke rising slowly from his keep, saw it thicken, grow darker, stronger, and heard, in the dawn silence, the crackle of flames. Impatiently hurling the babe aside, to be caught by loving arms before it reached earth, he spurred his horse up to the edge of the rock, preparing to gallop uphill to his threatened fortress. But a dark shape had been creeping ever nearer, slinking closer behind rocks, hidden in low bushes, never observed, a cruel hunter seeking one prey. As the Red Lord reined in for a moment by the cliff edge, as he paused before action, now she sprang. Snarling with hate, the Fenris hound leapt high upon him, red blood shining in her one eye. Unbalanced, afraid, the horse neighed in terror. The close-bound foes hung together for a moment on the precipice edge, locked face to staring face in deadly combat. Then the turf gave way. The Red Lord vanished from sight, to fall through the clear air, beside high white cliffs and come at last, broken, to cold hell as he deserved.'

A slow, quiet hissing of spent breath greeted this ending. Beer mugs, left hovering in the air during the climax, were laid reverently down on the tables. 'That were some tale,' acknowledged Tom Waller generously. 'I reckon nobody can better that. . . . Here, I'll get thee some ale.' 'Nay I didn't like it, 'twere too ghastly for me,' said one old fellow. 'Aye,' replied his neighbour, 't' wife wouldn't've taken it. She'd have called out to stop.' ''Twere but a legend . . . a tale,' Tom interposed, 'don't take on about it. . . . It'll come to mind now and then i' bed.' He laughed out loud as one or two of the weaker brethren quailed visibly at the thought.

The thin stranger smiled at them from a tired face. It blurred in Joady's eyes, seeming to spread and fill the whole shaded room. His one mug had grown a dozen more, he felt his brain swirling with the thin man's images. The corpses of dead beasts rolled back and forward when he closed his eyes. If he went at once he might just get to bed. The roaring inside his head grew louder as he closed the bar door behind him, crossed the moonlit stableyard and climbed up the ladder into the hay.

IX

WORKING LEAD

Josh slipped in through the open door of his home, having seen the bear cub settled in its lodgings. His Mam looked up and sighed. 'Get to bed,' she said listlessly. 'Aye, Mam.' He went out to the tap in the yard, bent over for a drink, splashed some on his face, then climbed up the steep, narrow staircase to his close bedroom under the stone-slated roof. The small window was open as wide as could be, and it too let in the smells of the yard, sweet, heavy undertone of the privies, sour current of hot drains and grease mingled with occasional whiffs of sheer pleasure as the smoke from Bert Ridding's pipe drifted through the still air.

Josh undressed in the dark and slid, naked, into his own corner of the iron-framed bed, feeling the coarseness of the blanket against his skin. He threw it off and lay there sweating, tired, unable to sleep. He rolled over, hanging half out of bed, his face against the gritty plaster of the wall. He thought of the cool, running beck, still out there under the stars.

Sounds pierced the thin boards of the floor. His Mam was trying to get Gran to bed. 'Aye, you'll get the cream right enough . . .' He could hear the sulky voice, the mutterings in reply, and then the creaking on the stairs as Gran was pushed upwards. He remembered how his Mam would shudder sometimes at the thought of Gran dying in bed one night and waking up beside a cold corpse in the morning. 'And she'll be laughin' . . . I know she will . . . the old devil . . .'

A thin partition split the already small area of the upper house into two tiny cubicles each sharing half the window and with room only for a bed apiece and space to close the door. Every word, every movement, could be heard through the laths as Florrie somehow got Gran partly undressed, 'take off thi sleds . . . nay, I'll not do it for thee . . .' and left her lying on the bed, still clutching her bottle.

A little later the stairs creaked as Bert Ridding came to bed. He brought a candle with him, part of his own store, for he was a saving man. The light wavered over the boxlike space where Josh lay, partly covered by the blanket, his eyes wide open. Bert's text hung framed above the bed. 'We will drink no wine; for Jonadab, the son of Rechab our father, commanded us, saying, "Ye shall drink no wine, neither ye, nor your sons for ever." Therefore, thus saith the Lord of Hosts, the God of Israel: "Jonadab the son of Rechab, shall not want a man to stand before me for ever".' As a brother Rechabite Bert Ridding would never want in sickness or old age, for his half-crown weekly from the brethren, nor a fine elm coffin with brass handles when at last he went to his Maker. It was a comfort to him.

He looked at Josh and smiled, stooping over the candle flame so that his face, lit from below, flickered and changed like a fickle ghost. But Josh had no fear of Bert and relished the chance to talk with him quietly of an evening. 'Can't sleep . . . eh, lad? It's thy last holiday, eh? Work'll start soon enough, and tha'll be a good help to thy Mam.' 'Aye Mister Ridding . . . but I'm tired.' 'Well . . . sleep if thou can . . . I'll read a while.' He opened the door and bent down with the candle to drag out his tin box of possessions and find some books.

'I've not got a job yet, Mr Ridding.' Josh had been thinking things over as he lay awake. 'I don't think I'll get into Dewhurst's.' 'No, lad,' Bert grunted from his doubled-up position, 'they generally take on one or two half-timers, but you have to know someone. They're a stuck-up lot.' 'I don't fancy being a doffer or a piecer, or a reacher-in for that matter.

There must be other things to do.' 'Oh, there's other jobs, right enough,' Bert answered, 'the trouble is getting one.' Josh's anxieties ran deeper than this narrow choice of workplace but he had no skill to express them. 'I'll miss t' becks,' was all he could say as he kicked the blanket aside.

Bert was coughing as he straightened up and sat down on the bed to strip to his shirt. He built up the two pillows against the head rail of the bed and leant himself back, his thick muscular legs protruding from the shirt tails as he wriggled his toes and coughed again. Because of his bad breathing he had to sleep sitting up. He put the candlestick on his lap and peered forward at the book in his hands.

Josh watched the craggy face beside him, with the deep running scar, edged in purple and red blotches, the flesh smoothing into a depression below his ear like fat melting into a pan. It had been a horrid fascination to him as a small boy when his father had died and they had taken in Bert as a lodger.

Josh had never really known his father, save as a distant memory of some weak, slow-moving creature that lay groaning on a high bed in the undivided upper room. He had always lived with his Mam and Gran amid their constant bickering, and so Bert Ridding had slowly become a person more real, deeper and safer, than any other adult. Bert he respected as a serious thinker, who read every night, who never lost his temper, answered proper questions with due consideration and took thought, even prayer, before he acted. To Josh this was another, and better, world.

He lay calmer now, happy to watch Bert's face as he read. The deep contours, fissures almost, on either side of his nose, were weathered by age, like the rock he had worked on. Such a combination of harsh, battered exterior and inner strength lay in Bert's craggy, enduring face that Josh stared on, drawing comfort to face a shaky-bottomed world.

Bert felt the gaze. Slowly he turned to look at the boy, 'What's wi' thee, young man?' 'Nay . . . it's nowt.' Bert looked at him appraisingly, for Josh's gaze was fixed again on the scarred depression in Bert's neck, with its veinings of angry red and purple, its seas of ugly yellow, its puckered edges of dead white mountains. He stirred restlessly. 'Mister Ridding,' he asked cautiously. 'Ay, lad?' Bert laid down his book and wheezed softly as he waited for Josh's question. He had never dared ask before, for it all seemed too personal, too dangerous, as if asking how a cripple got his club foot. Josh had enough sense, even as a lad, to know when and where respect was due.

'How did it happen, Mister Ridding? Was it an accident?' 'What?' said Bert, reaching up with cautious fingers as if in memory of pain. 'This? Nay I don't mind it now. It stings now and then in t' sun and a touch of grease from t' lead still drives me mad till I've washed it off. . . .I'm surprised tha never asked before. It shows clear enough . . .' He reached under the bed and laid his book down on the tin box.

'Dost' a want to hear what lead's been to me? Gaffer Lead, I could call it, from the day I were ten years old to this very night. Lead's given me work enough, hard grafting for it, and brass for meat, though never much. Though, the Lord knows, there are folk worse off nor I've ever been. It's seeking lead gave me these muscles,' Bert looked down at his legs, 'and this hole in me neck. Lead brought me to love and lose a lass, and lead made me broken-winded, like an old horse, before I were forty, and right glad she kept free of me. I'd be shamed to see her pretty face growing grey beside me in this place and life.'

Wound up now and going strong, as he loosed the stream of memories, Bert eased himself back among the pillows and stared blankly at the dusty half-window, unconscious now of his audience, but reliving for himself the lost world of his past. 'They say I were born not five hours after Queen Victoria came to t' throne, though that were nowt to me. I've lived her subject all me life, but what did she know of me? Me Mam died when me brother were born. Me Dad were a lead miner himself, up on Grassington Moor, Duke of Devonshire's workings, and we lived in Grassington, no more'n eight mile from here, but it seems like a thousand. Grassington were a pretty, busy town in them days and now it's half empty.

'Any road, I started at ten, so that's well over fifty year I've worked for Gaffer Lead, and, most on it, for his Lordship the Duke. Ten year old I were when me Dad took me up on t' moor. Three mile uphill in rain to start work on t' dressing floor at six, but we generally got a lift on a cart in them days. That was when lead were twenty-four pound a ton and they were taking on new hands.' Bert's fingers curved tightly over the bedrail as if he gripped an imaginary dressing hammer.

'It weren't so bad out on t' dressing floor. We'd sit round on mats, us lads, and break rock from t' ore wi' spalling hammers on a big flat knocking stone, bruised all white with blows, before it went into t' crushing mill and then to t' hotching tubs to wash away t' rubbish. Ten-hour shifts we worked, but if there were a break in ore coming, then we'd lark about in the heather an' look for plover's eggs. That were in fine weather. In winter we were cold.'

For a moment he closed his eyes at the vision of that desolate landscape, dotted with surrealist constructions, stark towers, long runnel flues, spider shapes of wire ropeways, giant revolving wheels, in black solidity against the yellow white of frozen vegetation high above any gentleness, and he shivered involuntarily although his body still was sweating stickily in the closeness of Roger's Yard. Grassington Moor in winter rose before his inner eye, the broken sweep of dirty-white and sombre grey mill-stone country, from Hebden Gill, the lowest beck, almost a little river, where the arched tunnel entrances of drainage levels emerged from hollow underground, the scar of the Duke's new road sweeping over the Gill and on up the workings, towards the ribbon of metallic veins winding sixty fathoms underground, past giant Cupola Smelting Mill, its walls reddened with the heat of roasting ovens, where the long runways of flues, raised mole-runs in appearance, turf clad, strained like clawing fingers over moss and rock to the squat smelt chimney in the middle of the field. The water wheels, driven from gutters of channelled, brown, peat water, high, spidery, iron and wood monsters, at High Grinding Mill, Brake house, High Grinding Engine. Then, wierdest of all, the long wire fences of the winding ropes, taking power from the wheels, which had to be located where water could be harnessed, then striding across country over pulleys and turn-wheels humming and vibrating with power, to where that power was needed, the mining shafts and their loads of broken ore, Sarah, Old Moss, Cottingham. Higher still the black moor rose, not gracefully, but lumpily, clumsily, rough and ill-shaped, to the tarns and dams, full of black water, that fed the becks and drove the wheels; the Blea Beck Dams, Baycliffe Dam and, closest to the summit, ever wrapped in cloud, Porphyry Tarn and Priest's Tarn, their banks marked by the scattered rocky rubbish of old washing floors.

Bert drew in a heavy, laboured sigh. His eyes opened slowly. 'My, it were cold,' he repeated, in memory of the wind piercing his sheep's wool-stuffed ears and his finger ends too numb to feel the hammer handle, ten years old and a growing bow-legged lad, coming up to five foot in height. And then, as the mind will, he drifted onwards towards another winter, when the frost sealed the earth like a prison wall, tarns and becks froze into black ice and grey stone sluices became a forest of icy pillars. Fourteen years old, a full worker in the deep shafts. 'I were down in Sarah, chopping out rock on t' top floor, standing on wooden beams above the hole below, and slinging down t' bouse. I had me powley pick and me plug and feathers. The rock was good and soft and a good shoot of ore in it. Soon enough it were piling up all round, till word came to give up work, for no water, no power, and they couldn't lift buckets out of t' shaft. So we came up into t' frost. It were a grim sight, gangs of black miners, in hard felt hats like basins, with their little tin oil jugs fixed wi' wicks on t' front, trudging across t' crackling moor wi' picks on their shoulders. Grey white cloud above and all round, no sight of moor's end anywhere. Work was over, and no work meant no pay.

'Then they called for little uns. "Go on, Bert," said me Dad, "hurry on, lad, there's work for t' little uns." So I hurried on. They took us to Cupola Smelting Mill. All were shut down there, for there were no ore to smelt, but the stone round the ovens was still warm, so we sat down with our backs against it.

'There was a ganger in charge, a short, round-faced feller, not much taller than me, wi' a great bristling black beard streaked wi' grey. He were known as a poet. "Right, me fine fellows," he called us over and told us what to do. There are round-headed flues running from each oven for half a mile up

1a Barge unloading at the Low Mill on Springs Canal

1b Show decorations. Arch at Ship Corner, looking through to Caroline Square

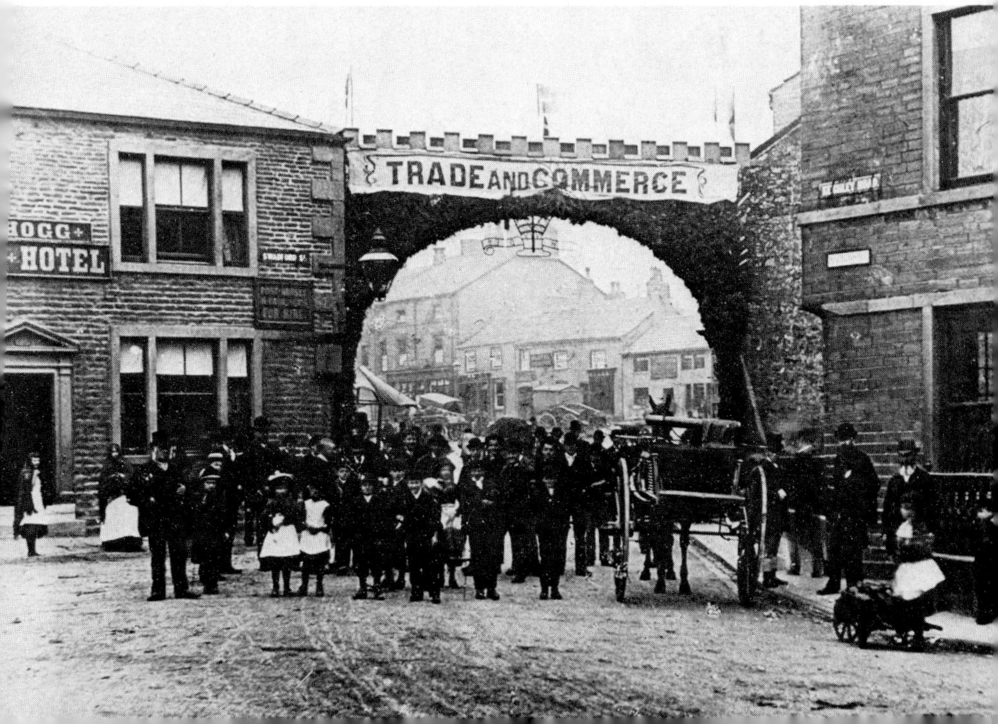

2a A wet market day in High Street

2b Skipton Castle. Gatehouse in foreground and Tudor wing behind

t' hill to a big chimney. They give a better draught and stop the fumes killing sheep. Now when the ovens are fired some of the lead mixes with the vapours that come off, and then it gathers on t' sides of t' flues. Now that the mill was shut down it was time to clean out and get it back. So they sent t' little uns in to scrape down the walls.

'It were like stepping into hell. The oven was still hot enough inside, and marked wi' fire, the stones red and charred so that pieces crumbled to powder in me hands. Any road, I clambered up to t' flue entrance. It were lower than me, and I had to stoop all the time, and the walls were veined, streaked wi' such colours, red lead and shining metal, patches of blue and green. All sorts of colours. I began to scrape it off, ready for gathering, and the dust filled the air and coated inside my throat till I were near retching all the while.

'So I stooped and scraped and coughed for an hour or two, wi' working uphill the while in a fog of red dust. The noise of me shovel, scraping over the rocks, grated through me teeth, but after a while I began to hear other noises, cracking like, creakings and shiftings. Sometimes they were near by, sometimes far and echoing down t' flue. I paid no heed, just banged away at the stones on the roof. There were a few loud cracks close by. Then, down it came on me! Cracked up wi' frost, t' broken keystone hit me hard on the shoulder like a great, big hand coming down. I fell to me knees and thought "this is it" but I kept t'shovel upright above me as the whole roof caved in. Down it came, slivers of stone, then lumps, blocks of roofing stone welded all together by frost. All went black and I waited to die.

'I woke and found meself frozen to a heap of stones, aching everywhere, twisted under and round like a spider's husk beneath a leaf. I wanted to cough, but the pain struck through me whenever I hauled in breath. So I lay, and a little cold light slipped in through the fall, and I thought now I would die. I were too weak to cry out, but I could see the metal scoop of the shovel and the red dust still clinging to it, so I lay still and breathed a little.'

Bert's whole body had grown rigid in the bed. He was sweating. Josh, moved by excitement and concern, had to speak. 'But you got out, Mister Ridding!' 'Eh?' Then, after a pause, 'Oh aye! I got out, or rather, they found me. But I lay there a while as the light went and I began to die slowly until they came. Yellow light from whale-oil lamps were in my eyes. "He must be hereabouts," they said, "there's been a fall." "Careful," said a deep, round voice, and I knew it were t' ganger, "don't bring it all down. Bring some props."

'"He's alive," said t' first one to touch me, and I remember thinking in surprise, am I really? For I didn't know, you see. It could all have been a dream, you see. Well, they carried me carefully down, and the pain were such that I slipped away time and again, but they always brought me round w' a nip o' brandy. I were young and fit then.

'Ganger took me to his own house, which were much nearer than ours, and that's how I met her. I saw this vision of golden hair, right golden it were, not fair, but rich and strong like . . . like the first turning of a fine frosty leaf fall, maple trees in the hedges. It hung over me and I heard the singing voice, soothing words. "Poor lad. Poor lad, what's happened to him." She were not more'n a year older than me.

'So she cared for me, and I didn't die. Though I were near enough to death for a week or two, so they couldn't dare to move me. Not that I wanted to move, for that were my first holiday from work and I lay like a baby and smiled at her face when it hung above me like the sun. She washed me down with warm water and cleaned off the dust, saying "poor lad, pretty lad", all the time and drawing in breath when she came on all the cuts and bruises. My legs lay all anyhow where the bones were broken. They sent a company doctor to see me next day. He did his best, but I grew no more in height after that.

'So I fell in love wi' my nurse. It were easy enough, for I had a soft heart for the lasses, and she shone over me like a bird wi' wings of gold. And she got fond of me. How could she not, even if only out of kindness, as if I was a lost puppy-dog or a sparrow with a broken wing, for she had a fountain of love inside her. And so it happened, her father were out working long hours, but she sang about the empty house and I lay quiet, listening in heaven. She smiled at me, I smiled to her. Then we talked a little. She put soft hands on my wounds and sighed. I, when I could move a little, reached out to touch her hair and she didn't refuse me. It weren't till t' day before I had to go that I found the courage to kiss her.'

Josh turned sideways in response to the depth of feeling that weighted Bert's simple words. There were tears in the old man's eyes. 'What happened to her, Mister Ridding?' he whispered, awed by this mystery, perhaps on the threshold himself of similar wonders, ready to dream of golden hair and warm, soft lips. 'What happened?' The tears were running down the old man's cheeks.

'Nay,' he burst out harshly, 'nay, it were better so!' His mouth clamped shut and he wiped his cheeks with an abrupt movement. 'I should have been stronger. . . . Even a lad of sixteen can follow his love. . . . We courted right happily for a year an more, courtin' like lads and lasses ye know, wi' daisy chains, ribbons, little songs and a kiss now and then. But we meant it, and both knew we meant it. Her Dad saw, but he were a poet and he'd have us sitting together on t' settle of a Sunday evening to listen to what verses had sprung inside his head on t' moor. I wish I could remember 'em now, for they brought tears to our eyes, all three.

'Then one day I heard t' news and ran back to be there before it reached her. I ran like a daft thing, straight across and down, over heather, down old hushes, right over walls in one jump, and I got there just as he did. There were blood coming from his mouth, and all the white in his beard were stained red. They said he'd been working inside Old Moss, way up from adit-level – that's drainage way – seeing to t' fixing of new stopes. It's like a great thin cave there, you work a ribbon vein along its shoot and up as well, so as you get higher you plant great baulks of wood across to give a new floor and up you go. By the time you've worked a year or two it's like a cathedral, high, dark, an with only hat-lamps shining here and there up in t' roof where there's ore to be won, and the shadows of wooden framing falling down on drainage pools at level. Any road, they were fixing a new platform, a stope, and he went down wi' a baulk that slipped.

'That were that. There were no more poetry for him. He were all crushed inside and died that night. She cried on my shoulder and two days later her aunt came to take her away. She were carried off to Saltaire, down by Bradford, and my life felt empty.

'It were the aunt, sure enough, not just t' distance over yonder. She were a woman of hard holiness, she had her hair strained right back off her face and were the sort that's always parading other folks' sins. My letters were sent back. I got little from her, and that posted secretly. So I went over. It's a long road to walk, there and back on a Sunday. Every time I were turned away. Then th' aunt said "I'll send for t' police". Sure enough, next time I came there were a policeman walking along t' street. I weren't going to run away, but I walked on without ringing the bell. My heart ached. I sought to die.

'It were no use my going, that I could see. 'T would only upset her to see me waiting outside. So I stayed at home and grieved. Sometimes I dreamt I were a knight on a big horse riding off to save her and win my love, like in the stories. But the world's a harder place than fairy-tales. Ye can't keep free of love. At least, I reckon them that try just shut out the flavour of life, you can see it in their scraped and empty faces. But, Lord knows, to lose it once it's been yours leaves a hole bigger than Old Moss in your head and heart. To hear she were married off didn't seal it, neither. But time passes even so . . .' his voice trailed away into a momentary silence.

'Nay, I'm talking a lot o' nonsense. . . . Pay no heed, lad, live thy life for thiself. Now get ye to sleep. . . .' Bert was obviously discomforted by all the revelations he had given way to. Josh looked at his worn face with fresh respect. 'Please don't stop, Mister Ridding,' he said, 'what happened next?'

'Well, there's not much to tell. I worked. I kept my pride, for my sake and for her memory. I worked and slaved, I joined t' Rechabites and found some fellowship there. I read some poetry now and then. I were a good grafter. Years passed by and everyone on t' Moor knew me. They said I'd be a ganger myself soon enough. After twenty years at work a man knows his job or he don't.

'It were about that time that cheap lead began to come in from America and Australia. Prices fell down to under ten pound a ton. There were nowt any of us could do. Wages went down. Then, at last, they said it weren't economic to go on, and they closed the moor down. No work, no pay, and Grassington were left half empty. Any road I still clung to lead, it were part of me, and when there were nowt else to do, I came over here, to Skipton, and found work at Fell's leadworks . . .'

'What about her,' Josh asked sleepily. 'Nay lad, there's nowt much to tell. I tried to forget. Enquiry only brought me pain. Though I did hear, not so long ago, that she's free now. Her husband died.' 'Are you going to see her?' Bert was silent for a moment. 'I thought I might try,' he said to himself, 'but I'm just an old romantic.'

Josh could hear his breathing slowing down as he recovered from speaking. 'Now, go to sleep, lad,' he said, 'it's a big day in t' morning.' Josh did as he was told. He drifted into deep unconsciousness and his inner vision expanded to mingle shining fountains of lead with golden hair like soft-feathered wings.

X

NIGHT PASSES

All Skipton slept. The poor in the dense yards about High Street, shopkeepers above their shops, the middle-classes in Middletown, the canal boatmen in their cabins and the rich in their big servant-filled houses scattered here and there in pleasant places throughout the town. Joady Thacker snored above the hay in the Red Lion stables. The main body of fairground people lay close by in Jerry Croft in tents and caravans. Giles the clerk was in his lodgings close by the newly built Christ Church. Baldisaro Porri and his wife slept unsoundly above their china after a second awakening by outrageous navvies who had chanted a final serenade outside the shop on their way back to Charlie's doss-house in Union Square. The navvies slept drunkenly now, but in Charlie's cellar the bear cub sat, wide-awake, with feathers on his muzzle.

Far down the Keighley road, at the very limit of the new, industrial Skipton, a single block of textile sheds, faced by a line of terraced houses, fronted the roadway. In one of those houses the two girls that Josh had seen outside the Coffee Tavern, Belle and Annie Tawney, had fallen asleep at last, but only after intense excitement had worn them down, for it was the Show tomorrow, and their big brother Jim was coming home from the war. Mother had a letter. It was true, and Jim wasn't hurt at all. The girls were put to sleep in the bathroom, the tinned tub hanging above their heads on the wall, for the house, though a good one, had only two bedrooms. John Tawney, their father, still known in Skipton as John Shaw, for hadn't his father come over from Ireland under a cloud and with an assumed name, sat up in respectable pillow conversation with his wife, Isabella. 'We're doing well,' he said, and his strong, black curling hair brushed to and fro against the cotton pillowcase propped behind him. 'Good,' she replied, 'you've worked hard setting up the dyeworks and we've still the girls to bring on.' 'Yes,' he mused on, 'we took a risk, a big risk, leaving Walton's Mill and setting up on our own. But we've got a good name for ourselves, there's none can do better sulphur-black, not at cop-dyeing. I'm looking forward to the Show tomorrow. I'll have a word with old Dewhurst. He's still council chairman and there's talk of naming me councillor.' 'Oh John, that would be grand . . . it's a recognition of all you've done,' 'Yes, I suppose it is. Well, say a prayer to St Teresa. Ask a blessing and we'll go to sleep.' He turned out the lamp and the last glow of bedroom light in that tail-end of Skipton died away.

In the big houses, among the Dewhursts, the Waltons, Scott the brewers and all the other manufacturing families, the servants had been hard at work on preparations all week. Horses were groomed, carriages cleaned down and polished, silk toppers brushed to an electric shine. Maids had worked all the children's clothes to a glory of stiff frilliness. The starched petticoats with goffered edges, the carefully ironed ribbons on the straw hats, the sailor suits with white, lacy collars were now laid ready for the morning's excitement. Mrs Scott, much put out by her husband's insistent worrying about money, had laid down George Moore's *Esther Waters* and sent the maid out into the back yard area to fetch more coal, for she felt chilly with sitting still so long. 'The brewery can't possibly be losing money, dear,' she said, 'look how large it is. We must put on a good face for the Show.' Mr Scott sighed and rested his head on his hands.

In Aireville House, the red-brick, mock-Tudor palace of the Dewhursts, overlooking the great mill, but still sheltered among pleasure grounds and pastures, old John Bonny

Dewhurst was unable to sleep. He dozed beside a small fire in the smoking room. The moulded plaster ceiling was stained yellow with tobacco smoke. He was making plans for the future, dreaming of new investments and wondering how he could keep his sons out of them, his mind occasionally slipping irritatingly back to the days when he worked for his father and the mill was still building.

In the Castle, as midnight approached, Barrett had completed his last tour of the sheeted rooms, a tour that always ended in the State Chamber at the top of the octagon tower, with its tapestries of the Siege of Troy and the tortures of the Spanish Inquisition, thumb screws, iron belts and strapados, the ghastly figures with eyeballs hanging down their faces. It was fortunate he was not an imaginative man, for at times he entered that room on a winter's night, alone, by lantern-light, and the pictures seemed to dance in circles round him. Young Lord Hothfield refused to sleep there. His bed-chamber was tapestried with Solomon giving judgment; a pleasanter theme.

High above the Castle gatehouse the monumental stone letters of the Clifford motto were picked out by moonlight. 'Desormais', they said, 'Henceforth', we shall keep watch over you for all time. The bulk of the Castle loomed over the town as it had for eight hundred years and Barrett, the guardian of its honour, went upstairs to bed.

Night passed away and the first traces of dawn found Badger stirring from a restless sleep. Silence lay inside the veil of mist that hung over the canal, a dreamlike absence of accustomed sound, so still that one might almost hear the smooth-shelled clams gasping out their watery breath on the muddy bottom.

Badger squeezed himself out of the cabin, climbing with rigid slowness, for the pain in his neck seemed to freeze his whole body. He put his hands on hips, scratched his low forehead and bent, groaning, to examine the damage. He saw cracked planks on the boat's waterline. A patch of chalico coating, the mixture of cowhair, dry horse dung and Stockholm tar they boiled up in iron kettles to waterproof and protect the elm planks, had peeled away and the lighter, greyish brown of the timber was showing. Badger shook his head in desperation, then yelled at the consequence. He was in for it now. Gaffer Bateman'd really be after him. Maybe he'd get away with just a row, but he'd be lucky. Badger's mind slowly sought to plan an escape or at least an excuse. He'd tie her up tight and pump out secretly if he could, then he could get away to the Show and say someone ran into her after she was moored. Yes, that might do. Meanwhile, there was nothing for it but to mend that split before it got worse. He couldn't risk the pull into Skipton. They'd take on tons of water. He must fit the stopping planks.

These were bent elm boards, pre-curved to the shape of the bow, kept in grease and ready fixed with rows of large screws, the points projecting a quarter of an inch.

First Badger had to clear out the cabin of other bits and pieces, spare ropes, tarpaulins, running blocks, box-mast extensions, and the sacking where the lad slept, so revealing the loose-planked floor. He brought up the oil lamp from its bracket in his own cabin and looked down into the slimy, foul-smelling well below. Water was still oozing in from a couple of damaged planks. He cursed and lowered his bulk into the cramped, sticky hole, positioning the lamp to his side, and sliding the greased boards in over his shoulder.

Fixing the boards was no great problem, though the effort of getting leverage to drive the fat screws into the tough elm drove Badger into gasping efforts and a mounting temper. The strain of keeping his head bent sideways sent throbbing pains from ear to shoulder. At last it was done and he told the lad to bring him a handful of picked oakum. Cursing the greyish, dusty fibred stuff, he drove it into the remaining crevices between the planks with a blunt chisel and as it felt the water it began to swell. Hardly anything was coming in now. Badger raised his streaked, weary face from the hole, wedged his massive arms above the cross pieces and kicked his way out.

Perse was still pumping lethargically. Badger lurched towards him. His legs were half asleep, his head was cocked sideways for ever, it felt, and his neck gave a twinge every time he moved, however carefully. 'Get that beast harnessed . . . before I kill thee,' he grated through his teeth, wanting no one else about to see him in his misery. The boy fled.

XI

JASPER'S CAREER

Five o'clock in the morning. Joady Thacker lay deep in straw, snoring the dust out of his flattened nose, securely lodged above the stables of the Red Lion. Once in a while he twitched and stirred, throwing the straw about, bubbling through his mouth. He always slept badly after a drinking do. Now he stirred in the straw, dreaming of a gigantic mug of ale waving in some outstretched hand at the farthest depth of a cavernous, gloomy bar and a commanding voice that seemed to threaten as it invited, 'Come hither, Joady.'

Just as the mug shone temptingly at his finger ends, just as the voice seemed to bellow inducements in his very ears and his mouth opened in expectation, his body tensed and a dreadful sensation of falling gripped him. He yelled out loud. Then his breath was knocked out of his body. He glimpsed a flash of grey light through a spinning opening. All was still again. Slowly his eyelids drew apart. He was on a pile of muck and hay, that at least was familiar, and staring between the legs of disturbed cattle, hearing their rumbling complaints, while above him the framed hole to the loft seemed to be rocking to and fro. His side hurt where he had fallen. He rolled over and staggered, wheezing unhappily, into the empty yard to get a mouthful of water at the pump.

The water was deliciously cold and sweet, limestone water, an old friend. He splashed at his face, stretched and straightened, yawning as he looked around him. Nobody seemed to be stirring, save a pallid 'boots' who was lethargically dabbing at a vast pile of jumbled footgear from his uncomfortable perch on a broken barrel. 'Mornin,' said Joady. The boots ignored him. Joady shrugged, drawing his pride together. 'Better see to yon bull,' he said to himself in encouragement.

So Joady made his way across the jumble of Red Lion Yard to where Jasper, peering inquisitively over the top rail of his pen, greeted him with a muffled snort, a welcome that mingled sweet hay breath with Joady's smell of sour beer. Jasper shook his head. He quite put the shambling, blear-eyed figure of Joady to shame as he stood, wide-legged in the soft morning light, dappled rich strawberry on white, deep chested, breathing regularly, giving intermittent, thoughtful, snorts. Jasper knew something was in the air and his bulging eyes were alert. This was not like the long, silent mornings on the farm, with only the distant cries of curlews to distract him from his peaceful chewing. His eyes rolled a little whenever scents drifted into his nostrils and he shuffled in his pen.

Joady leaned over to scratch his rump. Jasper had always been a good-natured, sensible beast, nothing skittish or dangerous. 'Sithee, old fellow,' he said, 'I'll be back to give thee a good scratching . . . don't fret. I must clear my head.' Shaking the offending object cautiously, an ill-timed experiment that forced him to screw up his eyes, Joady splashed more water in his face, groaned and staggered out into the High Street.

Not a soul was stirring. The dark stone fronts of the buildings lowered at each other under a shining grey canopy of sky. The flag of St George hung limply above the church tower. It was going to be another hot day. Joady followed the paved walkway from Town Hall to Churchyard and took a sheltered seat where a curved wall projected down the street like the prow of a ship. From this vantage point he could see everything of importance, provided he opened his eyes.

Only empty symbols of the coming Show and holiday, the collective excitement that would envelop Skipton, were to be seen, a few scattered objects in place of the throng of visitors. Half a dozen pens of sheep had been hodded up on the cobbles before the Black Horse Hotel and, as Joady closed his eyes, their patient bleating made him feel at home.

A random scattering of empty carts, battered wheels in the gutter, shaft ends resting half way up the slope of cobbles, stood beside partly-built stalls. Soon the stall-holders would be coming to display their produce – the eggs, butter, cheese, vegetables, oatbread and housewares of the Saturday market. But this was not an ordinary market day, with its usual single row of unpainted wooden stalls, grey and faded from the weather, that customarily lined High Street. There were strangers in town, curiosities everywhere, filling every gap in the line, each paying an extra twopence stallage to the frontagers in the shop and inns behind. They extended the customary range beyond the Town Hall as far as the Castle Gatehouse and the road to the Show field, a sudden, flaring growth of invaders, like yellow dandelions on a sober lawn. Painted, gaudy, red or yellow, some encrusted with fragments of mirrored glass, surmounted by furled awnings of striped canvas, or sloping, boarded roofs decorated with crudely-painted scenes, the show stalls added a new element of wild, irresponsible life to the quiet market place. 'That's a fine do,' Joady thought, promising himself a saunter past the display once Jasper was fully groomed.

The sun was already beginning to burn through the mist, and the rows of newly planted lime-trees in their tall, iron guards cast long, faint shadows in parallel lines across the roadway. Joady took the scene in with careful movements of his head. 'Nowt's stirring,' he thought and, resting his cheek on the dew-soaked stones of the green, northern side of the churchyard wall, he closed his eyes again.

Time slid by until a doubled sound of squeaking stirred him to consciousness again. One long-drawn, low-pitched squeak at regular intervals was accompanied unexpectedly by a chorus of higher squeaks, starting suddenly and tailing off with warning. Joady peered over the wall. The heaving figure coming up the slope from Mill Bridge was long-armed and broad-shouldered, bullet head pushed forward as he trundled along a four-handled sack trolley. A three foot half-cask sat on the trolley, cut from a big beer barrel, partly covered by a piece of sacking. The whole contraption rocked unsteadily as he came up the road.

At the top, just below the churchyard wall, the fellow stopped to wipe his brow. Then the long squeaking note of a badly fixed wheel died away and a chorus of squeals rose to a crescendo. 'Must have a litter o' pigs in that barrel,' thought Joady, and wrinkled his brow, for he hadn't much time for pigs, they were too much like folk, noisy, greedy, pushing, never letting you alone. The man trundled his hand-cart forward a short way, then halted above the sheep pens. He rested the trolley on its legs and edged the half barrel over. It rocked to and fro for a moment before dropping suddenly to the ground with a thud. The yelps and squeals were redoubled, and a circle of small snouts rose above the rim in noisy protest. Joady chuckled at the sight, thinking how like they were to the upturned faces of the choir-boys in church when the parson clipped an ear or two for whispering. The man threw the sacking back over his livestock and, after stretching himself, wandered up to the churchyard to take a seat near Joady.

'Fine morning,' he grunted. 'Aye,' replied Joady, 'it'll be hot.' The man made no reply. His social duties were over and he slumped in silence on his bench. Joady looked around for another subject of conversation. 'Not many folk about . . .' he said, tentatively. The man grunted. Joady's eye rested on an old green tombstone. There were few new headstones in the graveyard and hardly a sign of recently disturbed earth. An air of peace had attracted him, the old yews overhanging and the long, restful outline of the old church. 'They don't die very often here, I reckon.' 'Nobbut once,' the pig porter muttered laconically in reply before resuming his slump, chin on knuckles. Joady gave up. His conversational powers were exhausted. He felt embarjassed by his failure and laid his forehead on the cool stone once more.

Whether the outburst of squeaking had done it, though the piglets were silent now, or whether some other, inaudible, alarm had sounded, was by no means clear, but the High Street now sprang to life and demanded Joady's full attention. Shadows stood out clear and strong in the sunlight. New carts rattled in from every street. Small groups of cattle were driven spasmodically on to the cobbles and tethered outside the array of inns that doubled with shops as the main constituent of High Street life. A farmer's lady came round the corner, leading two bony, unpromising farm horses with tired eyes, and took up lazy stance in the churchyard, holding the halters negligently as the horses waited patiently on the roadway below.

An Irishman – it must be an Irishman, for his soft felt hat hung low over his brow and unkempt projecting hair – was driving a small flock of six geese along the cobbles in search of a tethering place. He saw Joady's face over the wall, raised a well-thumbed portion of his hat brim and said, with natural courtesy, 'Good-mornin' to ye, sir.' Joady responded as he would to any fellow beast, with a disarming smile and heart-felt, 'And good morning to thee.' The Irishman tethered his geese to the church gate and sat down beside them on the steps, drawing out a cloth-wrapped hunk of wheaten bread from his pocket.

The tired geese subsided in a heap, pecking feebly at the scanty handful of grain he threw to them from another pocket. Their feet were scarred black and pink where the skin showed through a grey film of road dust. 'Them birds need shoeing,' remarked Joady conversationally. The Irishman finished chewing a mouthful. 'Indeed,' he replied, 'from Heysham docks they've walked with me, and it's a long road.' He sighed, rubbing his knees. 'Three times I drove the lot of 'em through melted pitch and then through sand. Just to build up a good shoe, you understand. But it's broken away on the way.' Joady nodded in sympathetic understanding. The Irishman lay back on the steps and went to sleep, hat over his face. The geese slept while they could.

Joady sat watching for a while, and his eyes brightened. It was almost time for his breakfast. He chuckled to himself once again when he saw two lads, a couple he thought he recognized, one large and dark haired, the other thin and wiry, with a stubble of fair hair, slip out of a narrow entry between two shops carrying a bucket and a stool. Joady was grinning already at the sight of the larger lad, for he was enveloped in a pair of breeches much too large for him. They must be his father's and every now and then he gathered the loose flapping fabric around his shanks, glaring round in embarrassed fear of an audience. With ostentatious secrecy they crept carefully towards a group of tethered cows left unattended outside the Hole in the Wall, whilst their owner had his breakfast. 'Reckon I know what they're up to,' thought Joady as the lads peered all round like a pair of supremely cautious burglars. Then, as the big one kept watch, the little one sat hastily down on his stool and began to work a cow's teats. Joady could hear, in his mind, the soft hiss of milk into the bucket and smell the warm, creamy liquid. His fingers pulsed in sympathy with the movement. But, sharply now, the lads were off, ducking under a cow's belly and on to the next. The thin lad was quick and clever and in five minutes they were trotting back with a half-filled bucket, chattering and laughing to disappear once more into the darkness.

Joady rose, still amused. 'The cheeky beggers,' he said as he stretched, shook his head and explored his eye sockets with tender fingers. Then he strolled slowly and luxuriously across to the Red Lion. He had time enough and to spare for everything. Mr Stockdale would be in by ten, say, allowing for a good run in the cart. Jasper must be groomed and got into show fettle, but that was all. Joady scratched his head in satisfaction and squeezed through into the bar. He was enjoying his outing to the Show and all his anxieties about big city life were forgotten.

Only by a few seconds did Joady miss seeing the two mischievous lads emerge once more from their secret entry laden, this time, with a much odder set of instruments. A length of string hung over the big lad's shoulder, while each of them clutched under his jacket a rattling bundle of empty baking-soda tins, square cocoa tins, Australian canned-fruit tins, that jerked precariously in their arms as they ran. They crossed the road and sidled down the lane beside the Red Lion. Then they crossed the empty yard and bent down by Jasper's pen. 'Sithee bull,' said the smaller one, 'just keep still!'

Joady failed to observe the two grinning faces that peered round the wall corner as he came out of the back door of the Red Lion and darted out of sight when they saw him. Joady ambled slowly, in pleasant anticipation, towards Jasper's pen. He was still chewing on a piece of buttered oatcake and had a scrubbing brush in his hand.

Jasper bellowed at him when he was still four paces away, an irritable, grunting bellow with a note of petulant surprise in it. Joady quickened his pace, thinking, 'What's up wi't old beggar?' Jasper bellowed again. His eyes were rolling. He stamped his hooves, the iron half shoes ringing on stones. At that moment a strange, metallic, rattling noise burst out behind him, like a clanking of hollow chains, and Jasper started forward.

Joady's breakfast had done him good, but his brain still felt like the inside of a haystack. He paused to scratch his head, muttering thinly at Jasper, 'Whoa boy . . . whoa,' in a soothing way. But Jasper was not to be pacified. He jerked his rump again and the weird noise followed. By the time Joady noticed and understood the significance of the string of cans fastened to his tail, it was too late.

Jasper surged forward with a bellow of rage and wildly rolling eyes. The flimsy pen was no match for his strength. Half-rotted posts snapped off. Rails were battered to the ground. Joady leapt frantically aside, spitting out chewed gobbets of oatcake, as Jasper careered by.

He charged through Red Lion Yard, out of the gate and into the lane. Behind him followed the echoing, clattering tins, strange and threatening, as if a devil, a predator from his ancestral past, were hard on his heels. He ran faster. He bucked and kicked. The pursuing monster only pressed him closer, with greater fury.

The lane was blocked by a knot of fairground people wheeling a trolley laden with painted horses and twisted, gilded poles and chatting idly as they coasted along. Jasper appeared, snorting. Grins were shattered, mouths dropped open. Some scrambled against the high walls, others pelted out into High Street, yelling 'Bull's out . . . bull's away . . .' The stack of painted equipment toppled slowly, gracefully, sideways on to Jasper, butting him, scraping his flanks, adding fresh irritation. Breaking into a desperate gallop he roared, head down, into the High Street.

The echo of his resounding bellow, a challenge to all the world, froze the busy folk in their tracks: stall-holders, early shoppers, innkeepers at their doors, drovers by their cattle and children playing in the dust. A stupendous silence followed, the calm before a storm. From the depth of Kendall's Yard a trumpet sounded in reply as the elephant, penned close in darkness, smelled the crisis and signalled her acknowledgement.

The storm broke. Adults and children scattered – running for life as Jasper charged – diving in heaps into shop doorways, scrambling over walls, crawling under carts, peering out apprehensively from behind tree guards as his mad career took him across the broad street in a swinging curve. Still the monster behind drove him on to more frantic efforts and the din resounded up and down from cobbles to roadway and back again.

Joady ran after him, left far behind, begging, pleading, perspiring and weeping. He paused for breath and flailed his arms in a vain attempt to distract the bull and recall him to his senses. 'Jasper! Jasper!'

Another swerve, another lunge with his back legs and the pig barrel was gone, rolling down the road, whilst a bundle of amazed piglets shot out into the sunlight. Squealing with fright they ran in every direction, spreading fresh havoc, entangling themselves in the feet of the farmers and farm men who were at last recovering from the first shock of Jasper's arrival and beginning to group themselves in a purposeful crowd, armed with sticks and ropes, for his capture.

It would not be easy, for he was thoroughly roused and kept at fever pitch by the invisible tormentor behind him. The dogs were out as well, yapping at his heels, adding to his irritation. Jasper's eyes rolled continually, frantically, as he galloped, and foam dripped from his muzzle. He bellowed time and again.

A little girl was out early, in clean grey dress, black knitted stockings and buttoned boots, coming after breakfast to watch the sights. Jasper saw her dress move as she skipped across the road. He charged. Men shouted. The girl stood and screamed. Joady buried his head in his hands. Then she ran, face staring in terror, still screaming. Jasper was close, a great, bulky monster with glaring eyes. Stubby, vicious horns threatened death. Jasper's snorting breath billowed round her. Would she be tossed and trampled? All seemed still in High Street save for that one intense point of impending tragedy. A higher shriek followed. Jasper thundered on. The onlookers gasped and ran forward. But the child was safe, snatched by some fast-moving youth from the bull's path and into Manby's doorway at the very last moment. Joady shivered in reaction.

As he galloped down the street Jasper was confronted by an opponent worthy of his anger. Perhaps he saw in it the source of that demonic noise which pursued him. Four-square in his path stood a wooden stall, flapping its canvas wings in challenge, so it seemed. Jasper paused to summon all his strength, scraped his hooves once, and charged.

He struck it hard, thrusting with his horns. Its angled framework tore at his flanks. He bellowed and charged again. It rocked, trembled, lurched backwards and toppled unresisting, a puny opponent. A dismal shriek rose from the doorway of a near-by shop and drifted down the street. 'My eggs . . . he's broke my eggs!' Jasper, with lowered head and stamping hooves, churned to and fro over his recumbent victim, smashing planks and boxes, ripping the canvas canopy, splattering yellow stains over the cobbles.

Then he was off again, for the pursuing sounds were not silenced, running down the street. Joady puffed behind him, despair on his face. The assembled farmers were following. In front of him the roadway cleared dramatically. Yellow slime trickled down his horns and dribbled over his face. He reached Caroline Square and found both the archway at Ship Corner and the end of Newmarket Street blocked with the carts of fleeing people. There he hesitated. The dreadful noise was still behind him. He saw a dark alleyway and a flash of sunlight beyond. Down he charged, still seeking refuge, and the noise followed him across the flags of Birtwistle's Yard, clanking louder in the enclosed space so that he was cantering in full panic when he reached the open again and suddenly found water at his feet.

With a mighty splash and a sheet of rainbow spray Jasper landed in the middle of Waller Hill Beck. He swung downstream, slowing as he walked on the loose stones of the streambed. No sounds followed him. It was as if his pursuer had suddenly drowned. Stone embankments hemmed him in on either side, and when he came to a low stone bridge he stopped. The silence, the sound of running water, had driven away his furious excitement. He snorted, lowered his head and drank. A trickle of the yellow slime slid from his horns and slipped away downstream.

XII

GRAND ENTRANCE

'Must be late, nigh on nine o'clock,' groaned Bert Ridding to himself from his heap of dusty pillows. The air in Roger's Yard was already close and heavy, stifling to his mucous-filled lungs. He coughed, spat into a rag and slowly levered himself upright. He had slept badly all night. Then, with a body's ironical disregard for the hopes and expectations of its inhabitant spirit, as dawn came, bringing coolness and a welcome mist, heavy sleep sank him down, was still half with him yet so that he felt blear-eyed, aching-browed. What a way to start a holiday.

Bert eased himself over the edge of the bed and folded back the clothes in his tin chest. Carefully he set aside the darned and scrubbed shirts, verifying his inventory as he unearthed the best collar and cuffs he could find. He wouldn't trust his clothes to Florrie Bargh, for he might never see them again, and they went regularly to be laundered with a neighbour. His twenty-year-old best jacket was wrapped in yellowing, ragged newspaper. He had had it made when he thought, just for a little while, of a girl who might fill his life, a woman with dark hair who had stirred his imagination only for it to die away again, coming to nothing. But for some odd, unconscious reason he always re-wrapped it after its occasional airings in the same relics of newspaper, like a long-coffined shroud. The cloth was still thick and unworn, brownish black with a good close weave, cut out by the tailor in Grassington and finished with a short, rounded tail.

Bert stroked it with a flattened finger and laid it carefully on the bed. He had a few fine things, his books, his bone-handled knives, his jacket. These he hoarded in his damp-and vermin-proof tin chest and brought them out with ceremony. Today he would wear his best jacket, and already, as he struggled into trousers and stockings, the dense, collected heat of stone box houses in stone alleyways, crammed close together, was beginning to make him sweat. He worked himself into his jacket and stood up. You had to wear your best clothes for a holiday.

A shriek from the next bedroom startled him for a moment before his shoulders sagged in recognition. Through the lath partition that separated the cubicles the Barghs seemed even closer than they really were, as if they were shouting in his ear. 'Get up, yer old bag. Tha's not lyin there all day.' 'I'm not well . . . it's too hot . . . maybe it's cholera.' 'An I'm not getting some more o' that muck! If that's what yer think . . . Get yer dress on!' 'I can't . . .I'm too weak . . . I feel sick.' 'Then come down in yer vest and see if they'll let yer into t' Show.' 'Show . . . is it Show day? Why didn't 'a say? Been keeping it from me yonder . . . where's me best dress? Get me best dress!' The younger voice grew fiercer. 'Listen, old moucher, tha's got nobbut one dress. Dos't hear . . . t'other went long agone.' 'I want me dress . . .' 'Shut thy gob . . .' There was the sound of a slap and the old woman began sobbing, mouthing as she cried.

Bert Ridding, wishing nothing more than to get away and find himself a bite of breakfast, slid his boots on hurriedly and began to lace them up. Then he sat, head bowed for a moment, preparing body and spirit for the day, intoning within his head the short prayer he had learnt as a boy. From the next room came sounds of a breathless struggle, mumbled protests and sharp 'now thens' as Gran Bargh was got into her dress and apron. Bert rose cautiously to his feet. The floor boards creaked.

'Mister Ridding?' He sighed, it was Florrie's voice. 'Aye.' 'Is Josh there?' 'Nay, he must have gone out.' 'Drat the lad

. . . . just when I need him . . . Cans't give me a hand wi' Gran, please? I can't get her down on me own.' 'I suppose so.' Bert pushed his way into the other bedroom, past a mound of stained, faded blankets thrown over the bed end. The stench of old, unwashed flesh was almost enough to make him retch.

Gran's face was red-flaked, blotched with yellow patches. Her dress was dragged over the round, paunchy torso like a sack, half-full of soft meal. Yellow, shrunken legs, veined and dirty, stuck over the bed edge. She was maundering to herself and fumbling with spidery fingers in her apron pocket. Bert looked at her and a little sympathy surfaced above his accumulated disgust. 'Hadn't we better let her lie?' One baggy eyelid moved, and the flushed face shook. 'I'm going . . . I'm going to t' Show . . . . Over eighty year . . .' 'Shut up,' said Florrie, and the eyelid closed, but the hands still continued their furtive fumbling. 'She'll be t' death of me,' said Florrie, 'if she don't go to t' Show . . . come on,' and she snatched at Gran's legs, dragging her bodily across the mattress until her backside was at the corner by the door and she could be hauled upright.

Bert took her weight as they staggered downstairs. She was light enough, but even sourer smelling at close quarters. As he escaped to find some breakfast he marvelled at how long he had put up with living here. He had even felt obliged to promise to help Gran up to the Show. 'Our Josh'll be back to give thee a hand,' said Florrie. He felt a depression, a sense of inevitable degradation that no prayer could drive away. 'I've surely sunk low,' he thought to himself as he compared his present haunts with the bright hopes that had shone before him as a lad.

Just as he was leading the old woman out through the narrow crack at the end of Roger's Yard and into Sheep Street the lad ran into him. Josh was bubbling with laughter as he ran, his monkey face crinkled with lines and dimples as he paused for breath. When he realized that the way was blocked he stopped and grinned at them, hands on his hips, shirt hanging out of his breeches. The sight of such innocent, unburdened merriment brought a sudden easing to Bert's heart, an open acknowledgement that this lad was a consolation for him, evidence that God could always make a new beginning and save a lost world at the very end.

Bert leant Gran up against the shop corner and examined Josh with sardonic slowness, waiting for the irrepressible turbulence within him to subside. 'Well, young man,' he said at last, 'what mischief hast' been up to?' Josh's grin faded. He kicked with a clog against the paving stones. 'Nowt,' he said cautiously, looking up at Bert Ridding. Then the grin returned and Bert had to smile in response. 'There's such a to-do in t' High Street. . . . Come and look. There's a bull been out, and pigs running all over, and a stall full of eggs smashed.' He grabbed Bert's elbow to hurry him along. 'The Lord be praised,' said Bert, 'is no one hurt?' Josh stopped suddenly in his tracks. 'Nay, I never thought on that,' he said in an odd squeak. Then, echoing Bert, 'No, the Lord be praised, no one's hurt.' 'Well, help me with thy Gran and we'll go and see.'

As they emerged from the end of Sheep Street they found a scene of recovered good humour, of bustle, fresh visitors crowding in and rows of stalls open for business. In the half-hour since Jasper had vanished into Waller Hill Beck and Josh had followed to watch the fun, the world of Show and Town had blossomed into its morning of glory.

The sun was now high in the sky, hot on shoulders and heads, but not oppressive, a welcome gift of warmth, colour and life. It brought out a rich intermingling of smells, sooty walls and chimneys, dung and animal sweat from the roadway mixing with the stranger, menagerie odours drifting from Jerry Croft, the spicy, mouth-watering aromas from stalls, ginger-bread, brandy-snaps, hot peas and baked potatoes, and every now and then the whiffs of new hay from stableyards or the scent of a farm garden as a cart passed by, laden with flowers.

Traffic was almost at a standstill, a long line of vehicles backing all the way down High Street and out along the side roads, with a tangle of jerking, heaving horses at the crossing point just before the churchyard gate. Wheels grated together, ears twitched and the animals whinnied at each other, tossing their heads, whilst the red-faced drivers, the farmers, the coachmen in charge of parties of gentry, Johnnie Foster in his growler and the other professional cabmen, all were cheerful and tolerant of the ruck. This was part of the fun, and they waved ribbon-hung whips in greeting, shouting 'good-mornings' above the uproar.

The wide pavements were filled with a slow-moving procession of ordinary people, carrying baskets, babies, hats, bottles, blankets to sit on, cans of cold tea and flagons of ale. They came in hundreds, adding up to several thousand, leaving the new railway station, passing Dewhurst's huge mill, trudging over Belmont Bridge, then to run the gauntlet of all that proud array of Skipton's public houses before reaching the beer tent at the Show with its never failing river of ale supplied, as was proper, by Scott's Skipton Brewery. That was a thirsty day indeed, when road dust filled the throats of walkers and drivers, and temptation was ever close at hand.

This flood of drinking was a sad sight for Bert Ridding as the yeasty breath of jostling young men swam in his face. How often, in serious black-coated meetings, had he bewailed with his brethren the sins of Skipton town, its twenty inns for every single coffee tavern. So much resolution was needed just to keep weak men from degradation. But Josh was grinning wholeheartedly at every sight of the fair, the gawdy boxing booths with their fat, muscular, but oddly pallid champions, the penny shows, rich with glossy promise, fat women, living skeletons, dwarfs and giants.

There was a face Josh knew. He bounced up and down, waving his free arm to attract the attention of Bertie Whinsby, still active and unscarred by punishment, or so it seemed. He was now finely clad in a richly-napped jacket. His collar was tall, white and staring so that the lugubrious face seemed wedged on a newly sawn, wooden platform. This expression brightened to a pale smile when he saw Josh, and the pudding basin of a hat curled lower over his ears at every welcoming swivel of his stiffly held neck.

'Bertie,' said Josh as soon as they met, 'I see th'art still going t' Show. So am I.' He looked sideways and down at Bertie's breeches. Clean, pressed, not marked by any stain of mud or tear, a tribute to Mrs Whinsby's skill and effort. Josh winked. Bertie blushed.

Just the same face, larger, older, decorated with a pair of mutton-chop whiskers that had an artificial air, appeared above Bertie's shoulder. 'Come along, Bertie,' said his father. 'Oh Dad! I wanted to show Mop to Josh.' 'Well, hurry up then. Your mother's waiting. You know what that means.' Bertie hastily held up the slatted crate he had been cradling in his arms. It was almost filled by a big silky-furred rabbit. The long sandy ears hung out from behind the bars. Josh stroked one as if it were a velvet dress on a fine lady. 'She's beautiful,' he agreed. The large mild eye watched him anxiously.

'Come along,' said Mr Whinsby again. He knew that Josh Bargh was no longer proper company for his son. It was only a year or two since they had escaped from the yards themselves to newly built Brook Street overlooking the Dockyard, with its proper houses. Not more than ten years since his wife herself had left the mill. 'Brightest pair o' clogs on top floor,' they used to say. But that was all safely behind them and must remain so. She was well above it now.

'Do come, Bertie,' said a voice as mild as the rabbit's eye and Bertie jumped. 'Yes, Mam,' he answered guiltily. 'Call your mother, "Mother",' cautioned his father. Bertie went red, dropping his eyes to the rabbit. Mrs Whinsby shook her brown waving tresses in reproof, skipped lightly between husband and son, took an arm of each and, with a gay smile on her lips, a pair of dimples on her cheeks, said, 'Now you two men give me a promise. Shall we go to the Show directly?' 'Right away, love,' spoke Mr Whinsby. 'Yes, Mother,' agreed Bertie. A trio of backs, two large, clumsy, masculine, one pretty, willowy, feminine, drew away from that social contamination, Josh Bargh.

Josh rejoined Bert and his Gran, to follow the crowds up High Street where a row of stable lads lined the churchyard wall, their charges waiting restlessly below them, facing down the street, as if church and yard were some gigantic chariot about to be let off the brake and career uncontrollably downhill to the gas works. The wall beside the gates was now plastered with notices, placards in black angular lettering for all kinds of goods, amusements and exhibits, together with the extravagant display of a family of travelling ballad sellers. Their sales pitch was in full swing, gathering in its own little eddy from the main current of visitors.

'Come on,' said Josh and tugged at his Gran's quivering arm. As they turned and moved through the crowd the sudden sight of paradise shook the old woman into a staggering run which threatened to topple her over. 'Whoa there,' shouted Bert, but she was reckless. The stall was a small one, painted in a tasteful combination of yellow, lime green and pink, presided over by a flowering middle-aged lady in a brown print dress, with spreading rosy cheeks and a broad green straw hat decked in red and yellow flowers. The riches were displayed in piles, mounds of sticky, curling brown logs, the brandy-snaps of dreams, and at both corners the dishes of whipped cream, each in its bowl of cooling water, protected a little from the buzzing flies by a paper parasol.

Gran Bargh hovered among the flies, her mouth hanging wetly open, and scrabbled feverishly in her apron pocket. 'How much are they?' said Bert patiently. 'Penny apiece, love,' replied the stall-keeper, 'penny ha'penny wi' cream.' With a triumphant gasp Gran Bargh brought out a tightly folded wad of cotton waste from the bottom of her pocket and began to pinch it open. At length she drew out a silver threepenny bit, holding it tightly between thumb and forefinger as if it might fly away. 'I'll have two,' she cackled triumphantly, 'wi cream.' 'Right, love,' replied the flowery stall-keeper mechanically and deftly removed two sticky brown tubes from behind the piles, dipping them scrupulously into the cream at both ends. Gran Bargh watched with ferocious interest and snatched the treasures as they were handed over.

They stood by the stall, an incongruous trio, Bert in his best black, serious faced amidst the fun, young Josh restlessly alert to everything, his fair stubble almost white in the sunlight, rocking on his heels with impatience, and Gran herself, intent on her gluttonous pleasure. She sucked at the rolled sweets with her toothless mouth. Cream dribbled down her chin and on to the sagging bosom of her dirty dress. The angled sunlight displayed her unhealthy, blotched and wrinkled skin without compassion, but she was utterly oblivious, absorbed in a last remnant of bodily delight, one moment out of a handful in her long life.

At last she was finished and sucking her fingers. Her eyes were glazed, sunken, as if she had used up all her will-power in this desperate drive for sweetness. She swayed a little where she stood and clutched at the edge of the stall. 'Come on, Gran,' shouted Josh for the third or fourth time, and Gran began to stumble mechanically after him, as if nothing now mattered, for the best was over and she could leave her tired body to its own way.

They might as well have been soldiers of the garrison dragging back a dead comrade's body from some disastrous skirmish for all the help Gran gave them in getting her up on to the Castle Bailey. The jostling crowd flowed thickly along this broad rampart outside the walls, with more folk pressing from the roadway, which bore its own throng of wheeled traffic, up the curved stone steps, steeply cut into the bank, to fight their way towards the Show. From high amongst the Castle battlements a staring painted figurehead looked down on the scene below and the children pointed back to it, eyes shielded from the sun, asking, 'Mam, who's that up there?' 'Never mind, get along with you!' retorted their mothers, impatient at being held up by an old man and a boy dragging an old women between them.

They had to give Gran a breather when the pathway widened where road and wall reached level. Bert took out his pipe and looked up to the hills with a sense of longing. He didn't really like crowds, so many bodies packed together, as thick as lice in an old coat. He could see the whole craggy face of Rombald's Moor, the green of bracken, soft yellow-brown of dry grass and higher yet the darker brown of heather.

With hearing attuned to the cry of birds Bert was quicker than Josh to catch the distant, brassy notes winding round the Bailey from High Street, a ripple of music, weaving in and out of carts, beasts, carriages, some compulsive tune he had heard before, a march that led one irresistibly to follow it. This was the Skipton Mission Prize Band, led by their redoubtable conductor, Squire Firth, in his peaked cap and square moustache, soaring up the hill, inspired by the triumphal march from *Aïda* and carrying all the crowd marching with them.

The brassy rhythm came scalding hot from every instrument. Beads of sweat ran down the wrinkled noses of the cornet players to drop with an almost audible hiss, barely drowned by the row they were making already, on to the burnished metal of their instruments like spittle on a boiler. The parading crowds, now in a slow march as the band came through towards the Showfield entrance, caught up Bert and Josh, Gran sandwiched between them, and drew them onwards to the wide open gateway and the sight of tents blossoming above stone walls.

Josh pulled at Bert's sleeve. 'What's the matter?' asked Bert. 'It's no good, Mr Ridding,' answered Josh, 'we can't come in with thee, neither me nor Gran.' 'Why ever not, Josh?' 'No brass, of course. I never had none and Gran's spent hers.' Bert looked gloomy and fingered his hard-earned shillings in their thin purse. If he once started paying for the Barghs it would never end. He said nothing. 'Goodbye, Mister Ridding,' added Josh as he hauled Gran roughly out of the line. 'Where are t' going?' 'Hist, Mister Ridding, we'll see thee later!' Josh winked an enormous wink, grabbed his grandmother's elbow again and disappeared up the road as she tottered blankly behind him, making no complaint.

Josh and Gran made their way round to the back gate, the entrance for wheeled vehicles, horses and competitors. It lay down the back road to Embsay, through a small grassy courtyard lined by walls and barns from which a second gate opened into the main field below. Carts and carriages were already tethered thickly down the triangular field behind Storems Laithe, aligned against a row of young sycamores in dark green leaf.

Josh shoved Gran behind a wall corner and crept out himself to reconnoitre the enemy's defences. The high-wheels of carriages jerked, one pair after another, through the gateway as driver or coachman paid his entrance money. Two guards were there, stewards of the gate, armbands displaying their temporary authority, billycock hats a symbol of official respectability, even the brass waistcoat buttons a watchful, winking threat to any would-be illegal entrant.

Josh watched from behind a set of high yellow wheels, beautifully picked out in deep brown. The carriage drew slowly by and he had a brief glimpse of a pair of aristocratic profiles, two old gentlewomen surveying each other's faces as if that were not their customary entertainment across the whist-table every interminable winter's evening, as if the excitement outside held no interest for them. Josh slipped out behind and took hold of one of the back spring-frames, a half-formed idea in his mind. 'Whip behind!' shouted a passenger from the cart that followed. The liveried coachman cracked his long whip back along the coach side and Josh released his hold, to creep furtively away. In any case, he still had Gran to manage.

Perhaps it was as well that this attempt had failed at the outset, for as the carriage halted at the gateway and the coachman handed down some money, Josh observed yet more guardians converging in welcome on the carriage's occupants. A trio of the very highest officialdom, hung with scarlet pasteboard labels announcing that each of them was a vice-president, no less. At the carriage door they engaged in a triangular discussion.

'I insist that we should close the classes now, immediately. The time is decidedly past.' Mr Barrett had his watch open in his hand, his silk topper glistening in the sunlight and a short cane thrust under his arm. 'Well, Barrett,' retorted the impressively shaggy man who towered beside him in a high-crowned billycock, 'you know we always allow a little extra time, always have. It's only fair. Some of the people come a long way. They can't be precise to the minute.' 'Yes indeed, Walter,' burst in the Reverend Morris, his fine stomach and plump thighs squeezed into the costume of a prosperous gentleman farmer for this occasion. 'Indeed, I have only just been speaking, this very minute, to a competitor, an exhibitor I should say, whose bull has been delayed. He is most anxious, most anxious, and I had to reassure him. . . .' Mr Barrett shrugged and snapped shut his watch. 'Very well,' he said, 'another ten minutes. No more!'

The three vice-presidents were about to make their way

back into the main field, having saluted the Misses Hammond as they descended from their carriage, when a commotion broke out by the gate. Mr Barrett raised his cane vertically above his head, pinched his lips tightly together in the expression of one about to take command of a situation and said, predictably, 'What's going on here?' A chorus of humorous guffaws and shouts was his only reply. Some moving object had collected around it a gang of roughs, good-humouredly capering, concealing the focus of their attention from sight. 'Is that a late entry?' added Barrett. Another chorus of noisy comments was the reply, a broken chord that could hardly be divided into its component notes save for one long-drawn out 'It's flay-ay-aysome!' Then the roughs drew back in mock fear, screwing their faces, rolling eyes in simulated terror, so revealing the cause of this entertainment.

A thin voice filled the silence. 'It's only our bull, Jasper . . . he got out.' Even Barrett smiled shortly. Jasper was calm, docile, his mild eyes peacefully ignoring the noisy onlookers, but round his handsome face stood a sticky crown of hair, a tangled mane where the egg had stuck and run and set with dust and sweat in the heat of the day. Clearly Joady had done his best, for the bull was still damp from prolonged washing, but it would take more work with comb and bristle brush to get Jasper into show condition and no rubbing down would remove the broad graze along his flank.

'Well, well, well! He's a fine one indeed,' said the Reverend Morris with a beaming face. 'What do you make of his ruff, Walter?' Morrison looked on with twinkling eyes. 'An odd kind of lion,' he agreed. 'It raises grave doubts within me whether we have a proper class for such an animal.' Barrett looked evenly at the pair of them, took out his watch, tapped it, and put it back in his pocket. 'It's as well we decided to keep the gates open for late entries,' he added drily. The two stout gentlemen eyed him in surprise for a moment then, hands on hips, burst into a joint guffaw.

'There's Joady and our Jasper.' It was a woman's voice, emerging from under a broad straw hat above the side of a faded-blue cart that was still waiting close by the gate. The bull's ears pricked at a recognized sound, a blurred image of home, cool shippon heaped with hay, made him surge forward in hope. 'Look out, he's off . . . get back . . . mind this un . . . budge . . .' The roughs elbowed each other aside, fleeing from his horns in proper fear this time, though there was no malice in Jasper's mind. Barrett raised his stick as if to try the strength of his official position under such circumstances. Joady, hanging on to the nose rope, was dragged a few paces with heels scraping the earth. Half a minute of confusion followed as all eyes were on Jasper, whose reputation for ferocity was now established, and no one noticed, no steward or gatekeeper thought of checking the small, fair-haired lad who squeezed a limp, spidery old woman between a couple of hairy bargemen, dragged her furiously through the gateway and out sideways into the little triangular field by the gate, where he let go her hand and looked round for concealment.

The blue wheel of a farm cart rose above Josh's head as he paused to draw breath, to glance around him and see if he had been observed. A scent of flowers drifted down towards his nostrils, he heard a soft hissing, felt his hair moved as if by the lightest of breezes and looked upwards in sudden alarm to see a face watching him over the cart side. It was a pretty face, golden-haired under a bleached linen hat too big for it, with high cheekbones and a playful smile. One cheek rested sideways in coquetry against a posy fixed to the cart-rail. 'A girl,' thought Josh in instant response to his training as one of the lads, but the thought lacked the sharpness of habitual scorn. One dark blue eye winked enormously at him, the smile broadened and a teasing voice whispered, 'I saw thee.'

3a Manby's shop in Middle Row

3b High Street on market day

4a Old weavers' houses in Union Square

4b Lipton's shop in Sheep Street shortly after opening

XIII

BADGER'S HOMECOMING

While these opening moments of the Show were passing, one urgently awaited cargo had just reached the outskirts of Skipton. Badger was deep in slow mental convolutions as he prepared a story to justify himself to the gaffer. They came into town past the Craven Leadworks, under Pinder Bridge, keeping Christ Church tower right ahead until Belmont Bridge loomed before them and the horse began to draw up of its own accord. At last they had reached the Leeds and Liverpool Canal Company's yard, its high warehouses and steam cranes, situated close by the junction of the main canal with the Springs Canal that came from the Castle quarry.

Badger looked stiffly up and down the canal as he drew alongside and prepared to tie up. His muddy sense of disaster increased at the sight of an impeccably painted, newly washed boat, moored near by and covered in roses, scrolls, castles and hearts. He groaned inwardly as a lanky individual slid out of its cabin and stood, hands on hips, surveying his disgrace with obvious, sardonic pleasure.

'Had a good trip, then?' Badger ignored this sarcasm whilst he mounted the pump handle and began working up and down like a rigid-jointed automaton as he tried to avoid any motion of his fast ossifying neck. 'Hit summat again, eh?' Lot was watching every move as he sat, one leg crossed over his knee, on his cabin roof and plaited a rope end. 'I reckon that must have a fair load o' water on board,' he continued casually, as if speaking to himself. 'Mind you, maybe it'll give that boat its first real clean out for a year or two. My heart grieves to see it . . . 't were pretty as a picture beforehand. Does it stink?' Badger was stolidly pumping, resolved to ignore this mosquito voice whining round his ears.

'Hit summat, eh?' Lot was not yet finished. The mosquito was preparing to draw blood. 'Reckon tha'll need to pick a deal more oakum to plug it.' 'Shut thy bloody gob or I'll shut it for thee.' Badger was stung. He straightened up rapidly, yelped at the pain and looked round for some missile to throw. Lot giggled at his success. 'What's up wi' thee,' he added, 'hast' been bitten by thy horse, eh? The nag's revenge!' He raised a warning finger, straightening his legs and assuming a solemn expression as if he were Judge Hastings Ingham sitting in judgment.

'Lousy chunter . . . quit gurning. . . .' Badger grabbed his bucket from the cabin roof, thrust it under the pump and jabbed savagely at the handle. Lot grinned to himself, counting this a verbal defeat for Badger. The hold was emptying now and thicker slime, foul-smelling, was coming out of the nozzle. Badger took the full bucket, swung it idly in his hand, then jerked the contents rapidly, arching through the air, towards Lot's boat. 'Ouch, damn!' yelled Badger. He had forgotten his neck. Lot looked up. Too late. The spray of foul water struck his lovely boat, green slime splattered on the paintwork, smeared the whitened coils of rope and fell in large disgusting drops on Lot himself. He wiped his face on his forearm, then he glared across at the silently grinning Badger. 'I'll do thee for that,' he said. 'I owe thee summat already. . . . You wait yonder an' I'll get thee yet. Don't think to get away.' And Lot swung into his cabin taking his water can with him.

Badger looked warily around as he surreptitiously continued pumping in the hope of concealing evidence of his disaster. All the while he could feel an infuriating pain in his neck. Mayhap the gaffer would have left for the Show by now? But no, and his heart fell as a stocky figure strutted down to the waterside from the far end of the yard. Badger rubbed his suddenly sweating palms against his corduroys and was extremely busy in shouting at Perse to 'get on wi' it.' 'Where hast' a bin?' The gaffer was generally short with folk and this greeting was nothing special. 'I got held up at locks.' 'Slept late, I reckon. . . . I warned thee to get home in good time. Tha's too late! Hast touched owt?' he added, his voice rising in pitch as he changed from ritual criticism to serious concern. He leaned over the quay and peered closely at the bows, where the cracked planks could no longer be seen.

'Look on yon loading line . . . it's down. What tonnage didst' take on?' Badger shook uneasily and ran black fingers through black hair. 'About seven ton or so.' 'Where's t' ticket? Let's see . . . Aye, I thought as much. You've taken on nobbut five ton. There's two ton of water in t' hold, and more runnin' in every minute.' A sigh of resignation, with a good hint of threat to it, followed. 'Where did you hit?'

Badger, seeing a chance to shift the blame, began one of his half-prepared speeches. 'Them bloody bank rangers. It were a bank fall past Kildwick, and lad weren't lookin' out. I reckon he were dozing.' 'Were it dark, then? Who were't' helmsman? I told thee not to go on after dark.' Badger was silent. 'Right!' Gaffer Bateman shook his fist full in Badger's face. 'I'll dock thy pay for this. I should never have taken thee on. By rights I should make thee liver t' boat right now and get her out of t' water.' He had an air of grim amusement as he saw Badger's expression at this threat, for this would take until after Show time, and Badger would be filthy and exhausted by the effort.

'I've had Mr Morrison's engineer waiting all morning to collect that engine,' old Bateman added as he stumped back to the warehouse. He came out with the crane-driver and the engineer, Samson Yates, in a long, blue, sleeved apron which, if it had been white would have made him look like a shopkeeper, and a short peaked cap like a chauffeur's. 'I don't know what Mr Morrison'll be thinking,' he was saying, 'he wanted that tractor on display first thing. Now we'll be lucky to get her fettled and working by lunch time. You've put me in a right pickle. I should have gone over and driven her back. I told Mr Morrison at the time . . .'

'What's up wi' thee?' Bateman asked unsympathetically as Badger unlashed the cargo cover and rolled it back stiffly.

'Hurt me neck,' Badger replied defensively. 'Hurt thy neck! Come on, get on wi' it or I'll hurt summat else,' said Bateman, drawing a small flask of rum out of his pocket and taking a swig.

The crane engine was running smoothly. Pressure had built up in the boiler. Badger, Bateman and Sam Yates together saw to the lashings that would take the strain and hold the oil-engined tractor steady as they unshipped. Then Bateman gave the signal to hoist. 'She weighs nigh on five tons,' Sam advised, 'we don't want to get her swinging. Take it slowly.' So Walter Morrison's new toy was first displayed in Skipton and a crowd gathered on Belmont Bridge to watch her rise majestically in the air. The machine was like a miniature traction engine in appearance, with the same high, iron, spade-lugged rear wheels, the smaller front wheels steered by chains and a heavy flywheel at the side. She was painted a sober leaf green on the body, but the wheel spokes and bolt heads were picked out in red.

The crane began to edge her over towards the quay. It had taken them almost an hour to get such an awkward machine safely lashed and lifted clear. They were definitely short of time, but Samson Yates was still cautious. 'Slowly,' he warned, 'watch her, she's shifting a bit. Watch out!' A rope was slipping. The tractor canted over. 'Move it! Get her down!' Sam yelled to the crane driver. He reacted with speed. Down she came, dropping twice a man's height, landing with a crash on one of her rear wheels. She swayed on the angle of the outside lugs for a second, as if about to topple over, then settled back level. A cloud of fine dust powdered her paintwork.

'My God!' Sam was shouting as he ran over. 'What the blazes have you done? Can't even unship a piece of machinery.' He looked the tractor up and down. Bateman turned for a moment to scowl at Badger who was lurking beside his boat. Clearly it must be his ropework that was to blame. 'She looks all right,' said the gaffer, 'no scratches.' 'What do you mean?' Sam replied tetchily, 'it's the motor that worries me. That bang fair shook her up!'

'They're hard enough to start at the best of times,' he

continued more calmly, 'not like a steam engine, you know. There's no boiler at all, just this cylinder, and she runs off paraffin oil. Hold on a minute.' Sam went over to the warehouse where he had left his tools. 'It takes a blow-lamp, this job,' he muttered through clenched teeth as he struck a match and lit the flame. Turning it up high he then played the jet up and down the cylinder, trying the paraffin supply now and then to see if it was warm enough to vaporise and ignite. 'Takes some time an' all,' remarked Bateman. Sam nodded. The engine seemed about to fire. But the fourteen inch piston refused to move. He tried again and again until the blow-lamp ran out. Then at last he stepped back to wipe his face. 'It's no go,' he sighed, 'she won't budge. I'll have to take her down. I may be at it all day.' 'I'd best send word to Mr Morrison,' said Bateman. 'No, no,' Sam stopped him. 'Let's see how we get on first of all.'

Badger was still waiting. He needed his pay. Gaffer Bateman saw him. 'Well, th'art a fine, reliable kind of brass-head,' he remarked, 'what's wrong wi' a bit o' work?' Badger opened his mouth to speak. 'Go on to t' dockyard,' snapped the gaffer. 'Get tied up fast! Pump out right away, and then keep t' lad on t' pump. Do it thiself to clear t' hold, though. It wants getting out fast. Sithee – pay heed!' 'Aye,' mumbled Badger. 'Hast fixed t' stopping-planks?' 'Aye!' 'I want em well fixed, mind! There'll be no pay for thee until it's all done and damage assessed.' Badger groaned. He was relying on that week's pay for his day out at the Show. Cursing softly he untied the boat. Gaffer Bateman watched him. 'And get that horse to t' stables as soon as you land at t' dockyard,' he added, 'it's done enough today. I gave thee warning when I took thee on. Yon lad can take it while you start pumping!' Badger nodded and winced. 'Go on, get on!'

There was nothing to do but get the boat on to the dockyard, a level area of ground opposite Dewhurst's Mill. The dockyard was equipped with winches and cranes to haul empty boats from the water. There he resumed pumping and sweated in late-morning heat until at last he could straighten his back and go to the fore-cabin for a rest. He was back in Skipton at least. Somehow he'd get his brass from the gaffer.

Badger looked at himself in the fragment of mirror glass that leant against the side of the stove chimney.

Grease still marked his jowls. His hair stuck out everywhere in a dusty tangle. But his best clothes were laid out in the long locker above his bunk, the blue seaman's jersey and red neckerchief, the new corduroys lying under a lump of iron, and the deep plaited-string belt with its brass buckle. Badger smiled at himself. He'd show 'em all. None of 'em knew how he'd laid out his race winnings on a new outfit. He sighed, wiped his hands down his bulging vest, and prepared for pleasure. Sitting stolidly before the range, he dipped a scrubbing brush in the bucket and began to sleek down his unruly hair. He smiled for a while at the transfiguration. The best was yet to come. He was not aware that, as he rested before the mirror, Lot had taken a bucketful of water from the canal, crept stealthily along the towpath and, bent on revenge, was even now pouring it hurriedly down the top of his chimney.

XIV

IN THE SHOWFIELD

'What the hell have you done to my bull, Joady?' It was the voice of a justifiably angry Simon Stockdale, shouting from the driver's seat of the cart. Martha knew better than to interpose, when he began, at cross-purposes before he started, to try and get a full story out of Joady. All the while that Simon questioned his hopeless, despondent cowman, Sally was lifting loose petals from the posy of flowers. Lightly she held them in the palm of her hand, then opened her fingers to let them fall, one by one, over Josh's head and shoulders where they lodged in his stubble of hair like bright poppies in the barley. All this while he made no move to shake them away, but stood and watched and grinned absently.

'Mr Stockdale, there were nowt I could do.' 'Well, you damn fool, you blithering old collop. How do you expect us to show him like that,' and Simon pointed in disgust with his stick. 'I'll be getting a new cowman if you say owt . . . !' Martha laid a hand on his arm. Joady stood with lowered face, wondering in desperation if he would ever live this down and where else he could go. The gloomy silence ran on. To Martha the day felt ruined.

A short, harsh voice spoke up just behind them, as an ungainly, open-shirted, lantern-jawed man with two days' stubble running into the deep, black crevices in his cheeks prodded Jasper and Joady in the ribs one after the other with impartial professionalism. 'Judges won't like him,' he said to Simon, 'come along to me after . . . I'll give thee a fair price.' Simon nodded, still angry, but his wrath diverted to a new subject. 'Aye, Jemmy Rattle, I've heard of thy fair prices . . . Clear off and take thy offers elsewhere.' Then he turned to Martha, confidentially. 'But I don't know, there's little sense in keeping him after this . . . we've a good bull calf coming on . . . ' Jemmy ignored the angry words, looked as expressionlessly at the bull as if it were his own wife asking for a new dress, took his stick from his elbow and prodded its rump again. 'Bring him to me,' he said, 'tha knows where to find me. They all do. They all come to me. I'll see to him,' and he sauntered away, stick trailing behind him like a tail, towards the beast pens that were his natural home. There he knew every animal, could tell age, weight and price after one shrewd glance, though he would have been hard put to say who was prime minister.

'You'll not sell our Jasper!' protested Martha. 'Look at him,' replied Simon. 'I ought to sell 'em both.' Sally's attention was distracted from her game. 'Who was that dreadful man?' she asked her grandmother. 'Hush love! That were Jemmy Rattle. We were talking about him this morning.' 'Don't let him take Jasper. He looks a nasty man,' said Sally pleadingly as she threw a handful of petals towards the bull. Joady was already trudging away to the bull pens with his charge, full of shame.

Mr Barrett had returned to the gateway and now came over to the cart. Determined to see all problems resolved himself rather than leave them to the stewards, he tipped his hat to Martha and said, 'Morning, Madam. So that was the cause of all the trouble. Do you still want to enter him?' 'He's paid for,' replied Simon. 'Well the judges will soon be on their rounds. I don't know what they'll think.' 'The judges!' Martha sat up and straightened her back. 'Butter's still in there . . .' she pointed to the cloth-draped basket in the back. 'And it's hot! Come on Simon, we must get it in. I've worked on that butter.' Simon looked at her sardonically. 'Well,' he said, 'I thought tha'd forgotten it.' Martha returned the look. 'Tha were too busy talking wi' Jemmy Rattle to heed me,' she reproved him, 'and there's t' sheep to get in as well,' 'Aye,' agreed Simon, 'we'd best hurry.'

He pulled the horse round from the entrance. 'Steady!' shouted Barrett. 'What's the matter?' asked Simon, a little angrily. 'You're not allowed to take the cart any further. The parking is beyond the barn.' He pointed with his cane. 'We'll never get the butter in in time,' said Martha. 'Come on lass . . . we can only rush,' answered Simon, looking daggers at Barrett, for he hated to be pushed. Barrett shrugged his shoulders, replying cooly, 'Certainly Madam, get it in, get it in!'

But Josh had seen his chance. He jumped forward, leaving Gran shuffling alone in a withdrawn circle of her own fancies. He grasped the horse's cheek strap. 'I'll take her, sir,' he squeaked, 'I'll take t' cart.' Simon looked down in surprise. 'We could do that,' responded Martha, anxious for her butter. 'Come on Sally,' as she rose from her seat. Simon was more cautious. 'What'll you do with t' cart, lad?' 'I'll take it to t' parking place and wait there for thee, sir.' 'What's thy name?' 'Josh Bargh, sir.' 'Right, Josh Bargh, I'll trust thee. It'll be worth a threepenny bit to thee.' He handed over the reins and helped Martha and Sally down.

The sun was behind them and a brilliantly lit scene fell in gentle undulations from a clump of trees on the skyline. Below and to their right the treetops of Castle Woods rose from the deep ravine behind the Castle while now and then they had a glimpse of hard grey battlements through the leaves. Ranging uphill beyond a clump of sycamores and warty-trunked birches stood Storems Laithe, the old high barn where their cart would be waiting, and laid out in front of it were the cattle and sheep pens with their attendant show rings, conveniently placed between the competitors' entrance and the beer tent. Bellowing, bleating, muffled sounds of tramping hooves were a constant background to the hum of talk and excitement, the occasional shouts and laughter. Farmers, tall and short, raw boned or comfortably barrelled, all alike red-faced and red-necked with the day's heat, leant over hurdles, poked, observed, compared, sighed and scratched their noses.

Pride of place was given to the horse ring, lying in a hollow between two ridges, the Show Committee having decided to make cunning use of the contours and site the two stands on the sloping sides. Stand tickets being one or two shillings per person, this left the ordinary folk to crowd into the bottom and peer over each other's shoulders at the displays, the hurdle leaping and the series of parades that would reveal who were the prizewinners in each class.

Rows of white tents and marquees lined the level parts of the field, ranging down the wall side. There was a generous display of farm implements and supplies. Fred Manby was dashing about, cap half over his face, among his green painted chaff-cutters like giant mincers, equipped with pulleys for steam drive, at five pounds apiece. He surveyed the iron milk-kit carriers with long curved handle and high, thin-spoked wheels, the wringing machines, cake-mills, turnip cutters, and sighed with relief, for everything was still in order. The Stockdales walked past rows of iron-wheeled barrows, with a hasty detour round a display of Samuel Edwards' invincible patent horse-drawn lawn-mower as it was driven round and round on the same short patch of turf.

Sally savoured it all as they passed, knowing she would have more opportunities to explore later. It seemed a picture painted all for her, crowded with happy people, smelling hotly of canvas, crushed grass, sweat and flowers. But she retreated between her grandparents when they passed the monstrous beer tent, its banner flaunting Scott's Skipton ales, rearing up to the right of the main gate, larger than any other, the centre of a constant procession of pilgrims in and out.

The fancy stalls were ranged behind it, the retailers of knives, the pet food stall selling Old Calabar dog biscuits from Liverpool, the tailor from Keighley with a portable glass display-stand full of stiffly-clad dummies.

Baldisaro Porri, in his full glory now, with hair impossibly sleek, and blackened with grease for the occasion, middle parting, stiff collar and white cravat, was fast driving his assistants into a nervous frenzy. 'Do place these toilet sets so, Miss Calvert. They must be elegant. The price is seven shillings and sixpence, you know that?' 'Yes, I know that, sir.' Mr Porri gasped with momentary horror. 'Why have you placed these vulgar objects at the very front?' He pointed to a row of large, flowered chamber pots that seemed to stare every passer-by in the face with a reflective, highly-glazed immodesty. 'The farmers buy them. You told me last year.' 'Did I, did I? Yes, perhaps that is right.' Mr Porri flapped the tails of his coat and stalked behind the stall to check supplies. As the Stockdales went by Simon grinned at the chamber pots and nudged Martha. 'Be quiet,' said she, 'there's no time for silly remarks,' and she dragged him at last into the main exhibition tent.

They were only just in time, for the officials' and judges' tent was across the way and the judges were taking a glass of claret cup together before proceeding to the stern business of the day. Mr Anderton, the veterinary surgeon, came rushing into the crowd, reaching out for a glass as he cannoned into the stolid, muscular back of Mr Thomas Roose, head gardener to his Grace the Duke of Devonshire and judge this day of the root vegetables, swede turnips, kidney potatoes and long mangold wurzels. 'Sorry, Mr Roose,' said Mr Anderton, 'I've only just time for a gargle,' and he swallowed the glassful at one gulp. 'Nothing but late entries,' he gasped as he rushed off again, 'bulls, sheep, even rabbits . . .' Thomas Roose raised one bushy eyebrow in amazed response. The gravity of his position would never allow him to race about in that way.

Mr Robert Carr, a neighbour to the Duke's house at Bolton Abbey, and the chief cattle judge, was thoughtfully sucking the handle of his cane and eventually reached out with it to attract Mr Barrett's attention. 'I hear there's some trouble with the bulls, Barrett.' 'Well, we did have a runaway this morning, not here, in High Street, but no harm was done.' 'Thank goodness, thank goodness,' the Reverend Morris confirmed from beside Barrett's elbow. 'You shouldn't have any trouble with them now,' Barrett continued, 'I see your friend Robinson from Storiths has a prize bull entered!' 'Hmm,' replied Mr Carr as he resumed the sucking of his cane-handle. Barrett drew out his watch. 'Gentlemen,' he called, 'we are running late. Can you please begin at once.' Soon a stream of curly-brimmed bowlers began to flow from the tent to disperse in solitary droplets about the field.

The hutches for rabbits, fowls and pigeons were all arrayed in the shelter of an open-walled marquee beside the main exhibition tent. Bertie Whinsby had carefully laid his pretty Mop among the other rabbits, the silver greys and Dutch blues, lop-eared or straight, and then retired with his mother and father to enjoy the Show. He had already left when a late exhibitor rushed in and jammed the hutch of his heavy, fighting, Belgian buck within half an inch of Mop who, like her owner, was no fighter.

Bert Ridding, now in the field with his ticket bought, had not seen the Barghs, which was some relief, and had drifted over to the fowl-hutches. This was an automatic move, since his fellow-miners had all been known as rearers of prize fowl and so that was the place for any lead-miners to gather and meet old friends. A stocky old man rose from a squatting position by a hutch as Bert walked into the shade of the tent and opened his eyes wide to look around. 'Bert,' said the man. 'Mark,' was the reply. 'Middlin?' 'Middlin.' They shook hands firmly. 'Still makin setts?' 'Aye, still workin' at Fells?' 'Aye, I can't keep off lead, not now.' 'Always were a romantic, Bert, we all knew it. What with thy poetry an' all. Dost remember when we were lads together on t' dressing floor?'

Bert nodded. Mark raised a hand to wave. 'Look, here's Peter Thwaite and Bill Jackson. It reminds me of the days at Grassington Fair . . . before his lordship killed t' mines and the town with it.' 'It's no good being bitter,' Bert cautioned him. 'Well, here's th'owd softie,' said Peter Thwaite, slapping Bert on the shoulder, 'still dreaming, eh?' 'Don't talk so daft,' Bert replied, 'and don't forget, young man, I'd've been thy ganger by now if we were still up there.' Bert pointed in the general direction of Grassington Moor. At that moment the judges came in and Mark, who was showing a pair of old English game fowl with a bright red cockerel, retired to his hutch. The group of miners drew to one side to talk quietly of the old days when the world had seemed a promising place and each of them had a recognized skill and status in the community.

The judges were also working their way down the main exhibition tent amidst the piles of beautifully displayed entries laid out on long white tables. The tent smelt sweetly of crushed grass, newly scrubbed vegetables and ripe cheeses. Indeed, the local cheeses were already under scrutiny as slices were cut from flat dales' produce and crumbled thoughtfully into the mouth, while the two-handed screw-corers were driven into the heart of larger drums. There were dozens of entries for the butter prize and baskets of every shape and size, some lined with lettuce leaves, others still covered in clean white muslin, were ranged in rows three deep down the table.

Martha had added her contribution to the competition and her heart sank. She looked at another woman in a green print dress and large upper lip who was carefully re-arranging the pats in the basket. 'However are they going to decide?' Martha remarked. The woman ignored her. 'It'll not be easy,' Martha went on. The woman shook her head and turned away. 'Pay no heed, love,' a friendly looking old lady turned aside from her task of selecting out the best dozen from a hamper full of brown eggs. 'I know her, It's no good wasting words. She's that mean. If she had a mouthful o' gumboils she wouldn't give you one.' Simon, Martha and Sally all laughed at this. 'Come along,' said Simon, 'they'll be judging t' cattle. We'd best go and see what's happened to Jasper.'

XV

THE JUDGING

The judges had stopped at the first pen, a narrow space indeed and filled to bursting by a square-bodied, small-eyed, tiny-horned elephantine animal, blotched in chestnut brown on white. 'Short-horn bull, under eighteen months. Mr William Robinson, Storiths Farm, Bolton Abbey,' read Mr Carr from his list. 'Looks uncommonly like the Craven Heifer,' said Walter Morrison, who was keen to learn about the finer points of judging cattle, as he drew out a Craven Bank pound note from his pocket and compared the drawing of that celebrated beast with the animal beside him. 'Similar proportions and colour, quite remarkable, though this is a bull of course. Perhaps he has some of the strain.' 'Did the Craven Heifer ever calve?' asked Mr Carr. 'I don't know, but I'm told she was over ten foot in girth, quite a monster, like this one. A remarkable beast!' They moved along to the next pen. 'Here's the troublemaker.' 'Well, this is a fine specimen, a better shape than the last one even if he carries less meat, but he appears to have developed a ruff!' 'Apparently he attacked an egg stall in the High Street this morning. I must tell you the full story later.' Both judges chuckled as they moved on to other stalls.

Jasper watched them with a reproachful look. He could feel an atmosphere of derision and doom surrounding him like a prickly cloud. He snorted dismally. An unfamiliar creature wandered over to lean against his pen, a desolate lurching figure, like a spider suddenly exposed to sunlight. Jasper blew a welcoming snort at this fellow outcast through his flaring nostrils, but Gran Bargh ignored him and he was left to gaze blankly, meekly, at the bustle that passed him by.

There were a number of other bulls in the class, but Mr Carr spent little time on any of them. 'It's between the first two,' he pronounced decisively, 'let's have 'em in the ring and look over their points.' So Joady's hopes rose again as he led Jasper before the judges. 'Come on, old feller,' he whispered in the bull's ear, 'look bright. Shape up. We may do it yet.' Jasper shook his head.

The judges were going over their check-list of scoring items, looking for the broad head well-carried, full bright eyes, strong well-set neck, sloping shoulders and deep chest of the perfect bull. 'Can't see at all how you'll choose between them,' remarked Walter Morrison. 'No,' agreed Mr Carr, 'it's down to detail. One's got weight, t'other good build. Both move well enough. We'll have to run our hands over and check for damage, feel the fat as well. They're as full of breed and quality as an egg's full of meat.' He was warming to his task now and sucked his cane handle intensely for half a minute as he watched the two bulls shifting restlessly at each other's closeness.

'Walk 'em round again,' he said to Joady and to Robinson's cowman. He watched the bulls from this side and that, ran his hands along their flanks and down the muscular legs. At last he stepped over to Jasper, laying his stick across the animal's back. 'I don't mind the egg in his hair,' he said slowly as Joady shuffled in painful anticipation beside him, 'but,' here he paused. 'But there's a graze down his side. His skin's broken. It's not much, but this is a show. Otherwise I can't choose between 'em. I'll give this one second prize. He's a nice fellow, but he's marked.' He tied a purple card round Jasper's left horn. The monster at the other side received the coveted pink ticket and seemed to swell even larger in consequence. Joady sighed. It was not exactly a failure, but they'd hoped for victory from Jasper.

The judges had passed on to look at cows in calf when the

Stockdales reached the bull pens. Simon's broad back was stiff with righteous indignation. For some reason his womenfolk had become awkward and skittish, wanting to investigate china stalls and buy sweetmeats before the serious business of the beasts was over. Now he stumped ahead of them without waiting to see if he was being followed. Joady caught a glimpse of his figure some distance away, before his face was visible, and could tell Simon's humour from the way he moved. This was no place for a timid cowman, and Joady slipped away at once rather than face his master.

Simon could have no doubt who was the winner. A knot of jubilant farmers and cattlemen had gathered round the huge bull from Storiths Farm. In the pen next door a hot, slightly dusty Jasper had been hobbled, no doubt for fear that he would escape again.

Simon saw the raucous crowd at the next pen, heard their boasting and noticed the amused looks they were directing at Jasper. He halted and glared at the bull. A cloud of dusty flies exploded from a cow-pat as his boots squelched in it, and he slithered, almost falling. 'Damn!' he muttered, then, recovering his balance walked towards Jasper's pen. 'Damn the thick-skulled beast!' He clutched the rail, breathing heavily. 'Simon,' said Martha reproachfully, 'don't lose your temper.' 'I'm not losing my temper . . . !'

Grins appeared on the faces of the spectators at the next pen and Simon's mouth snapped shut, his face grim and hard. A man detached himself from the group and came over. It was Jemmy Rattle, the cattle-dealer. 'Well, Simon Stockdale,' he drawled, 'what d'yer say now? I told thee he'd never win.' 'He's a daft beast,' replied Simon as he scraped the dung off his best boots against a post, 'I've seen enough of him!' 'I'll give thee a good price.' 'Nay, I don't know.' 'Ye'll be a laughin' stock. D'ye want to drag him all t'way back to Littondale?' said the tempter casually.

Simon shook his head thoughtfully. He glared once more at Jasper and looked round to see if Joady was about so that he could vent his wrath on the sinner who was truly responsible, but Joady had almost reached safety in the beer tent. Simon kicked savagely at a post, the flies buzzed around his head, filling his thoughts with their insistent, sawing whine. 'Right,' he said, 'right! I might just do that. I might sell him now. What'll you offer? What did ye say he'd fetch?' Jemmy Rattle moved into bargaining stance, arms hanging beside his pockets.

Sally was almost ready to cry. Everything seemed to be going wrong on such a lovely day. She looked at the haggling men between the two pens. Her grandfather, transformed into some grumpy ogre, was standing at arm's length from this tattered devil of a cattle dealer with his thin bony shanks and the look of an executioner. Like a pair of prizefighters about to begin another round they held out hands warily, almost touching fingers. 'Make it nineteen pound!' 'Never, I'll give thee fifteen!' 'Fifteen, that's daft! Make it eighteen pound ten. He's a good beast.' Jemmy laughed sarcastically. 'Sixteen pound ten and shake hands on it.' He made to slap Simon's palm. Simon drew back. 'Nay, I'll take him home.' 'Tha'll never. Tell thee what. Seventeen pound and that's my limit!' He held out his hand, palm upwards.

Sally watched the decision growing on her grandfather's sweating face, saw his eyes turn to his own hand, the ridged and calloused surface, strong fingers and broken nails. She turned back to Jasper. The sound of hard palms slapping together, the words, 'Right – sold! sold!' shouted in her ears. 'A shilling back for luck!' added Jemmy Rattle hoarsely. 'I'll give thee a toffee! . . . Nay, go on, take sixpence,' replied Simon holding out the coin. Jemmy Rattle struck the hand once again, in anger at this meanness on the part of a simple seller, leaving the mark of the coin even as he took it. Simon sighed, shrugging his shoulders, his anger suddenly gone.

'Well, I hope you're pleased with yourself,' said Martha coldly, 'tha's upset our Sally.' 'Nay I'm sorry, rabbit,' answered Simon remorsefully, stroking his granddaughter's hair. 'I didn't mean to get rough like that. It's this heat. Seems to be threatening a bit.' Martha relented. 'Well, what's done's done . . . and we did get second prize even wi' Jasper in that shape. He'd 'ave won . . .' 'All right,' snapped Simon, 'don't go on about it.' He rattled the sovereigns in his pocket. 'Any road, it's nice to have some brass in hand.' Then a slow smile came over his face. 'You know, dealing wi' Jemmy Rattle brought to mind an old tale I once heard.' 'Is this one you've told before?' 'Nay, I think not, it comes from when I were a young man, from the days before they had newspapers up in the dales and so when a farmer lost a beast there was nowt he could do but pass its description around by word of mouth.'

Simon looked sideways at Martha with an inquiring smile. 'Come on, you'd best get it out,' she replied. 'It's only just sprung to mind after many a year,' he said, 'and the Lord only knows all the secrets of a man's mind. Reckon I heard it in a pub, and I don't go so oft these days, about once a year perhaps.' 'Let's have your tale,' said Martha.

'Right. Well, this farmer were well known as a tight-fisted old devil, and he'd a good milker gone stray, so he asked all over t' dale, seeking everywhere, not so much like a good shepherd as an avenging spirit. He poked his nose into other folks' shippons and he even bought drinks at a pub or two while he were asking for news.' 'You're inventing,' accused Martha. 'Aye, perhaps . . . any road, a week went by and he found nothing, so on Saturday he went complaining to the parson, saying as how he'd prayed to the Lord to bring back his lost beast and the Lord hadn't answered his prayer.' ''Twere no business of his to bother the parson,' said Martha. 'Nor the Lord neither, but the parson, being fairly new to the place, said he'd announce the cow's loss in church next day and ask anyone with knowledge of her to get in touch with the owner.

'Next morning came and they all went to church. The farmer, being an economical fellow, generally used church time to catch up on his sleep. He were still dreaming fondly of his lost cow when the parson began to go through the announcements. Well, he began, right enough, with the banns of a marriage, saving the cow till last, you see, and so, after he'd said "if any know of any just cause or impediment why these two should not be wed" he waited a moment, like they do.' Martha nodded in understanding. 'Well just at that point the old farmer woke up. He heard the silence, knew somehow there'd been a question asked, had his lost cow right in his mind, nothing else, and so he got up on his feet before the whole congregation, ‘Aye,’ he said, right loud, ‘her paps is all covered i’ warts and she’s warbles all over her backside.’

XVI

PORT AND ALE

Simon's joke had not gone down well with his family, and so he rattled his sovereigns for consolation as they wandered away from the bull pens. The morning was almost spent. The judges had completed their deliberations and retired to their shady tent for another glass of claret cup. A noon lethargy had fallen over Storems field. Dense, heavy blocks of heat seemed to have congealed over the whole tented town and the sun beat down on close-packed, strolling crowds from a brassy sky. Simon and Martha followed the rest, idly noting the wares, displays, the differing people in their brown billycocks, straw gardens, print dresses, boots, shoes and clogs. Sally's head was drooping as she lagged behind, her fair hair blotched with the heat.

Suddenly Simon stopped beside a stall selling cooked meats and lemonade. 'My belly's rumbling,' he said, 'it must be long into dinner time . . . and . . . why . . . hell's a poppin! We've left that lad wi' t' cart for more than two hours.' 'Aye,' replied Martha, 'we'd best get back, our Sally's been walking too long in this heat. Makes you sweat like a new cheese. Come along love, we'll go and sit down,' and she took Sally's hand. Simon led them uphill past a collapsed pile of bandsmen and instruments stacked under the blessed shade of the sycamores to the shambling collection of carts, wagons and carriages that were parked beyond Storems Laithe.

'Cart's there any road,' said Simon thankfully, 'and t' horse. But where's that lad?' Then he chuckled, nudged Martha and pointed. The early excitement, the holiday atmosphere and the heat had been too much for Josh. He lay, sprawled out in the shade of the cart, one hand still clutching the horse's reins. His thin limbs were twisted about in a surprising confusion of joints and baggy clothes and his quick, sharp face had relaxed, reverting to the defencelessness of a childhood he had almost lost. Martha's sympathy was roused. 'He looks thin,' she said, 'wake him up, Dad, and find out who he is. . . . We could spare him a bite.'

Simon got hold of the reins and gave a tug, then another. Josh stirred a little and groaned. In sudden surprise, as his eyes opened, he saw the underside of the cart, shook his head and scrambled to his feet, swaying a little in the hot air that oozed around him. 'Here it is, sir,' he squeaked, 'I've done nowt wrong.' Simon laughed sympathetically. 'Nay, that's all right, lad,' he said, 'but there's a hammer and peg under t' seat. Tha could have tethered her.' 'I didn't know.' 'Aye, well, what's thy name again?' 'Josh sir, Josh Bargh.' 'I see, and where dost live?' 'In Roger's Yard,' replied Josh rather shamefaced, for he was well aware of the reputation his homeland held in Skipton.

Simon pursed his lips and frowned. Then Martha stepped forward and took Josh's arm in her strong hand. 'Why he's thin as a lath, and we left him to look after t' horse for two hours or more. . . . What family has t'a?' 'Just me mam.' Martha nodded in understanding. It was very hard grafting for a woman on her own and often the children suffered, even in better places than Roger's Yard. Suddenly Josh's face took on a look of surprise, then he began to giggle, he couldn't help it, his face screwed up into a monkey's grin. His eyes watered. They all chuckled at the sight, part amused, part amazed. 'What is it, lad . . . what's up?' 'It's me Gran . . . I've lost her. I left her at t' gate . . .' and at the thought of his black, spidery Gran slithering blindly the round of Storems field, croaking bitterly 'Where's that dratted lad,' he fell into helpless laughter again.

The heat and emptiness fell heavily on him and he swayed,

leaning against the cart wheel, seeing the faded blue spokes blur before his eyes. 'That's enough, Simon,' said Martha sharply, 'let the lad sit down and eat with us. We'll have our dinner here under the cart, where it's shady. . . . And look, Sally here needs a rest and some dinner, she's as prickly with heat as . . . as you are.' Simon grinned. The point was taken. 'Well, if the Lord wills it,' he said, and swung up on to the cart to bring down two heavy baskets from under the seat.

Josh lay with his head resting bonily against the wheel, too tired to move it. Waves of heat and dizziness rolled over his body. His legs were strangely heavy. In a dream, a soft-edged shimmering haze, like the vision of the genie that Bert Ridding had read to him one winter's night, bringing jewels and strange foods in golden bowls to Aladdin and his mother, he watched the preparations for dinner. The baskets were unpacked, the creaking wickerwork revealing its treasures, fat brown bottles of ale and ginger beer, clean white glistening mugs, the thick crusty pile of new baked oatbread, a bowlful of half-melted butter, a pale, crumbly wedge of cheese, and with them lettuce, cucumber, radishes in a cloth. There were meats too, cold and tempting, a loglike section of meat loaf, pink slices of thick-cut ham, a jar of potted meat.

Overcoming his weakness in the glory of this feast, Josh swung himself round on an elbow and levered to sit upright. 'Come on, lass.' Martha's voice, which had been steadily reckoning over the provisions in the background, now broke into his daze. 'Come on, give the lad a sup of ginger beer. He'd be better off with that.' So, with a half-conscious grace of movement, her prickly heat having gone, Sally poured out a mug of ginger beer, grey and cool as limestone water, and handed it to Josh. He received it reverently like a knight accepting the chalice from his lady before setting out on crusade. Simon nudged Martha, grinning, and she shook her head.

The first sharp-edged, bubbling, refreshing taste of that ginger beer was like the conscious start of a new life to Josh. He sat up. He looked at the food ravenously. He noticed new delights brought out of the second basket, a bowl of red and white strawberries, a jug of thick yellow cream, and a cloth poke of sugar to pile on both. Reverting instantly to boyhood he sat up on his narrow haunches and held out a pair of hands almost like a begging dog.

Martha, seeing this, smiled the secret smile of a good housekeeper who is pleased to see her food well appreciated, and loaded a plate for him. He ate fast, drank in gulps, and after a while even had time to smile round at them all, at Martha broadly, at Simon cautiously, and at last at Sally confidingly.

This was the essence of a summer's day to Josh. Sally smiled and tipped her flower garlanded hat in his face. 'He does like butter, Granny. I think he loves butter.' Josh, too happy and well fed even to feel embarrassment at these intimate remarks from a girl, grinned in reply and delicately rescued an escaped blob of cream from a blade of grass, carrying it luxuriously to his lips.

The band had been playing a programme of steady, digestive lunchtime music for a while and now retired again while a thick silence flooded the whole scene with an almost tropical density of midday heat. The birds were quiet, animals dozed in their pens, prize pink and white pigs with their backs protected by a taupaulin from the threat of sunstroke. People were too hot to speak much and stretched out in the shade wherever it could be found. There was not a breath of wind. The party under Simon Stockdale's cart had become as somnolent as any.

Only desultory conversation was possible, but at length Simon turned to Josh. 'Didst' see our Jasper this morning?' Josh looked puzzled. 'Nay,' he replied, 'there were nowt but a bull out,' and he chuckled at the memory. 'It were a real comical sight, let me tell you . . .' 'That were our Jasper,' interjected Martha. 'Oh,' Josh's face fell. 'I didn't know. I'm right sorry . . .' Josh hung his head. Martha carried on the conversation thoughtfully. 'I wonder what's happened to Joady?'

She would have been surprised at the answer. A wandering cowman is a lonely creature when he comes among the crowded folk of this world, and Joady had been wandering aimlessly, drifting with the flow of humanity that ebbed and swelled along the grassy aisles.

No-one realized how deeply Joady had been hurt by the morning's disaster. A shy, indrawn, quietly stubborn man, his only resource being his skill with beasts, having a long-developed sense of understanding, an intuition as to their moods in relation to his own; he had felt sure of Jasper. That bull was a good beast, he was positive, not too wick, free from evil. Joady would have staked his life on Jasper's intrinsic goodness. In desperation he retreated to the beer tent. He could see no other refuge and, though he was not by custom a drinking man, he was led by the weight of past expectations to try and drown his sorrows in drink. The interior of the tent, once Joady had edged his tortuous way inside, apologizing as he went to fat bellies, dimpled elbows, boots and tent ropes as he was driven into them in turn by the swirling tides of drinkers, the interior resembled a reeking, resonant cave with translucent dirty walls filled with a steamy atmosphere that collected on the canvas roof and dripped on the thrashing crowd below.

Joady's daily life, alone, airy, spent with the beasts, gave him no immunity, no preparation against such claustrophobia. He worked around the jostling, shouting groups, towards the long trestled bar at the back of the tent. 'That'll be threepence,' said a yellow-faced serving girl handing him his pint of ale. 'Threepence!' mumbled Joady in astonishment, but he paid and crept into a corner of the tent, tight against the bar-end.

The ale gave him no pleasure. It tasted flat and warm, slightly sour to his tongue, coating his teeth. He sighed and drank some more. There was nothing else to do except fight his misery or savour it. Something jogged his elbow and the beer slurped up into his face. He spluttered, wiping his eyes with his sleeve. 'Sorry, mate,' came close to his ear, a deep, coarse voice of somebody already well entrenched at the bar. Through beer-stuck eyelashes Joady had a glimpse of his neighbour, of a wide, hairy torso partially clad in a black-stained vest speckled unevenly as if with soot, massive, bulging arms and shoulders, brown to the elbow, pallid above, and a barrel-like paunch supporting the grey-streaked corduroy trousers fastened with a plaited string belt. The head was small in contrast, like a hedgehog on a tree stump, little black eyes, upturned nose, low brow and tightly curled hair, whilst a graining of some grimy substance ran along the edge of the hairline and faded into the eye sockets.

The fat man raised his glass in a gesture of welcome. He seemed in need of company, a fellow in misery. 'Sorry, mate,' he said again, 'I were slownin . . . my, it's throng in here. . . .' He paused to survey Joady, the breeches, open shirt collar and baggy jacket. 'Art' a farmer?' he asked. Joady wiped his face again and nodded cautiously. 'Cowman,' he murmured. 'Eh?' the fat man shouted back. 'Cowman,' said Joady, a little louder, and the fat man nodded. 'Clap it down there,' he said, holding out a wedge of hand to be shaken, 'they call me Badger . . . I were christened Albert, but call me Badger. I'm a boatman.'

'Never trust a boatman, never do it . . .' he continued. Joady's eyes widened at this self-imposed insult, but Badger was wound up with his grievance and determined to expound it to this neutral audience. 'Right snod, some on 'em are, snod as a serpent . . . but I'll get 'em, I'll get 'im. Yer want to watch out, I tell yer.' Joady nodded apprehensively at this impressive warning. Badger had clenched his fist and was swinging it up and down against the canvas beside Joady's head, leaving a fresh smear with every movement.

'Look at me, just look at me . . . !' Badger's grief and self-pity were growing with every mouthful of ale. 'Wouldn't reckon, tha would never reckon that this were a new kit, bran new, woulds t'a?' Joady shook his head. 'Bucketful o' water,' hissed Badger cryptically. 'It's that Lot, an' I'll get him yet. Get him. I'll break him into fifty thousand pieces. If I catch him wi' Izzie! Bloody soot! Blaster, witherin, thungin, cronkin soot. I'll give him soot!' Joady was unable to retreat. His pounding head was pressed back against the tent. He could feel the condensed fumes soaking his back hair. His own misery was quite enough to fill his life to overflowing. Badger ended this clenched peroration with a harsh, threatening laugh, looked again at Joady's suffering face, interpreted it as sympathy, being bound up in his own disaster, held out his palm and repeated 'Clap it down there, mate!'

Joady's sinewy, long-fingered hand was folded into the fat man's enormous palm. They stood, close-pressed together for a moment in the silent falsehood of miserable good-fellowship. 'Give us thy mug,' said Badger, taking the half empty pot from Joady's unresisting fingers. One small, black eye winked suddenly, as with the speed and agility of one accustomed to working in confined spaces, he swung round with his back to the bar and bent down into a crouched position. His thick arm disappeared under the trestles and Joady had a brief glimpse of a tilted bottle letting out a stream of blood red liquid before Badger was standing before him, belly pressing his elbow, mouthing a grin and thrusting a full mug into his hand.

'What is it,' asked Joady, peering into the scummy brown depths. 'Hist,' replied his new friend, heaving up his trousers as he spoke, 'hist . . . that's best port wine. The very best. Drink it up, mate.' Joady found the combination of ale and port wine both repulsive and effective. He hadn't the heart to reject this well-intentioned offering and so he drank, and as he drank he felt more thirsty. His tongue seemed to swell in his mouth. His stomach felt hollow. While they drank Badger repeated his stealthy refilling operation, so the strength of the mixture gradually increased. It became sweeter, heavier, stronger, and the tent began to balloon inside and outside Joady's head.

At this moment a harsh, incoherent voice called 'Badger' over the shoulder of the boatman and a lugubrious face, made mostly out of badly-shaped bones, or so it seemed, slipped between them. The new arrival was wearing a mouldy tail-coat and bore an equally ancient silk hat in his arms but appeared to have no vestige of a shirt, nor, for all that Joady could see, of stockings inside his gaping boots. 'Why, it's Billy Gellin,' said Badger. 'Come on Billy, have a drop o' port wine. Hast been kebbin?' Billy, who earned his scanty living by kebbing, that is dredging the canals for lost coal with a perforated bucket on the end of a pole, winked slowly in response and emitted a grating noise which seemed to indicate agreement with Badger's proposal.

The port-inspired beer drinking began again. Badger retold

his grievances and Billy Gellin acknowledged his interest with renewed gratings. After a while Joady began to feel as if he were breathing in a warm brown fog. He slumped back against the canvas.

'Summat t' eat . . . a bait . . . come on. . . .' He felt the sensation of being dragged through the tent, the surprise of fresher air and sun full on his face. His vision cleared for a moment and he found himself supported beside a cooking stall, which boasted a portable iron range, a tall back chimney and a trembling heat haze worthy of a lead mill. 'What'll yer have?' asked both Badger and the stall proprietor. Muzzy-headed, queasy-stomached, Joady could think of nothing but his favourite winter food at the farm, his rare treat, to be savoured and anticipated with relish. 'A pork chop,' he groaned, 'a good thick un.' The stallkeeper raised his eyebrows, but Badger nodded and so he slapped a thick slab of pink pork into his frying pan. Badger eyed it hungrily, and sighed to himself. It was Badger who carried the plate away with almost proprietorial care and sat Joady down beside the beer tent. The sun beat on them. Joady, his head throbbing, closed his eyes and swayed where he sat. The sun pressed on his skull through his thin hair. He laid the chop down on his plate, rolled over sideways, groaned and fell heavily, suddenly, into the depths of unconsciousness.

Badger's face showed no expression of surprise as he surveyed this collapsed, ungainly, sprawling acquaintance at his side. 'First stroke of luck I've had all bloody day,' he was thinking to himself, his eyes on the unbitten chop now resting securely on a tuft of grass beside Joady's hand. But just as he began to move towards it a knot of blue-sweatered boatmen emerged from the beer tent and passed close by. One of them saw him. 'Hey, look, here's Badger. . . . What's he up to? It'll be no good, I reckon . . . .' 'That's right, that's right.' Badger ignored them. 'Been picking owt up, any more oakum?' shouted an elderly fellow from a safe distance. 'Nay,' replied one of the others on Badger's behalf, 'he'll not beat any more beasts' brains in for a while, not since he were in gaol . . . but mind he don't beat thine! Should see the look on his face. He's not gurnin, I can tell thee.' If Badger could have found a weapon to hand he would have thrown it, but all that came to his grasp was the pork chop. It clung to his fingers with a natural affinity. 'Bugger off,' he growled.

XVII

FESTIVE LUNCHEON

Elaborate cooking was not a skill easily practised in the new kitchens of Skipton Castle, the sixteenth century ones. Months ago the committee had decided that the festive luncheon, ceremonial heart of the Show day, when officers and members of the Society dined with their respectable fellow townsmen, should be held in the Baronial Hall and not in the tent. They had too many memories of sodden food doled out under damp canvas. The consequences of this decision were experienced by the cooks as they crammed pan after pan on to the iron range, bringing into service the carved-stone, charcoal stove under the mullioned window, chasing assistants hastily off to the New Ship Inn for ready prepared supplies, working at a dangerous pace in the heat. They were without exception buxom ladies with good, red faces, since Skipton was no believer in male chefs of the French style.

The heaps of food began to build up on rickety trestle tables, the dirty pans to collect on the vast, shallow sinkstone in the window enclosure of the preserving room with its single brass faucet. Hot, filling, pudding-like meaty food for a hot day, the meat loaves, long and solid as a yule log, the deep-crusted game pies, saddle of venison with red-currant jelly, the ham and beef in chunks, the special dish of cooked lobster for the top table, shining red on a bed of cabbage leaves. To accompany this solidity with rival weight came bowls and dishes of vegetables, steaming, pale, boiled potatoes, with young, tender carrots and sticks of green celery, a concession to coolness.

The puddings were waiting in angles, corridors, corners of watchtowers. There would be tipsy cake and trifles, cold rice pudding, fruit pies, raspberry and red-currant, all to be soaked with jugs of good, heavy, yellow cream. Then there were local cheeses to follow. The barrels of Skipton ale were stacked along the cellar passageway, under the round arched ceiling. Bottles of claret and hock for the more sophisticated were stored underground in the wine cellar, whilst for the ladies, in recognition of weakness, bottled ginger beer and lemonade cooled amidst mounded ice blocks.

The feast was waiting. The cooks, celebrating that blessed moment of peace when the preparation is over and the guests have not yet arrived to initiate the panic rush of serving, were fanning themselves with cabbage leaves.

A kitchen lad had been stationed as sole watchman among the battlements to give warning of the impending invasion. Lunch at half a crown a head was a guarantee of selectivity and soon the lad caught sight of the first outward signs that the successful and prosperous were on their way. A flower-like cluster of broad, lace-edged hats surrounded by a phalanx of glossy toppers was drifting down the Bailey-walk below him. Slowly they escaped from the white, blazing outer-world into the sheltered gardens of the Castle.

Conduit Court was cool and refreshing after the glaring openness of town and Show. The yew tree cast a damp and protective shade over cobbled pavings. Moss and green mould flourished on the stone seat below it. These first arrivals wiped their foreheads, shrugged shoulders to let the cool air penetrate their heavy jackets and began to perk up at the thought of luncheon to come, at the sight of moulded preparations and the beaming faces of the cooks seen through the open windows of the kitchens.

John Tawney, in the blackest of jackets, for a dyemaster could afford no tinge of blue or brown to mar his commercial reputation, decided to wait in the courtyard and try to catch a word or two with the principal guests before lunch. His wife

Isabella was looking after the children at the Show. It was a pity not to bring her, but at least Jim, limping but safely home from the war, could share the occasion with him. As they waited, more lunchers crowded past them to climb the staircase to the hall above and soon the officials began to arrive.

'Well, what have we here? Young Tawney, isn't it?' Walter Morrison held out a welcoming, bear-like hand. He was always quick to recognize and encourage. 'Welcome home, young man. Not so hot as the veldt is Skipton, but quite hot enough.' 'Yes, sir.' 'Kitchener's doing well out there. Did you read his latest proclamation?' 'Yes, sir,' shifting position to ease his injured left foot. 'I heard it read, sir.' 'Good, good. I made them send Kitchener out there, you know. He's a fine general, a great man!' Walter Morrison beamed down at Jim Tawney, and began to cross-examine him in detail about the true state of the army in South Africa.

John Tawney stepped aside. He had seen John Bonny Dewhurst coming and took advantage of this opportunity to have a word with the celebrated manufacturer about Skipton Town Council and the need to fill it with sound, commercial men. They chatted confidentially for five minutes, but it was enough. The opportunity had been well taken and John Tawney contentedly approached the coming meal. He found his son and hurried upstairs to claim a seat.

Sunlight danced on cobwebs through the panes of glass set into the roof of the high-raftered hall. Small doorways led to hidden, angular rooms at every corner. The walls were whitewashed. The boarded floor had been scrubbed to a damp whiteness with sand and soft soap. The long trestle tables were laid with linen cloths, An air of thick-walled coolness created appetite even in ladies who had vowed they wouldn't touch a morsel, not a crumb, my dear, on such a baking day.

John Tawney searched for a spare seat beside some useful neighbours, perhaps a customer or potential customer, but the hall was almost full. His son had been captured by a pair of young ladies, out to discover a hero, and John smiled to himself in amusement. He could rely on Jim not to let his head be turned. In the end he had to take whatever seat he could find, a vacant spot squeezed between a group of shopkeepers and the party of Mr Scott, the brewer, who was at the head of the table.

He looked around: young Fred Manby had been induced to take a glass of wine with Baldisaro Porri and was blushing at the public exposure this entailed. Mr Porri himself was in the best of spirits, had already proposed a toast to the ladies around him, had scornfully wiped the cheap, willow-patterned crockery with his napkin and was now watching the preparations for serving luncheon with undisguised satisfaction.

Mr Scott, who looked considerably more intellectual than a brewer of forty should, was rapidly disposing of a bottle of the best hock and declaiming the villainies of Nonconformists and Catholics to the Churchmen on the top table. He appeared to have forgotten his own financial discomforts in this opportunity of airing a High Churchman's views on the social and religious divisions of the town. As John Tawney turned an ear in this direction, Mrs Scott was already full flown in support. '. . . And I really feel that the Canon is so gentlemanly in his gaiters. A clergyman must be a gentleman as I regularly tell the servants at morning prayers . . .'

The seats at the high table had filled with judges, officials and important guests, and now Mr Barrett raised his glass. 'Well, gentlemen, we are in favour today, I think.' He pointed to the sunshine above, and acknowledgements to their good fortune rose around him. John Bonny Dewhurst alone said nothing, but fingered his white beard and waited in acid patience for the self-congratulations to die down. The Reverend Morris, invited to say grace, obliged and Mr Barrett signalled for the luncheon to begin.

The heaps of substantial food appeared, steamed for a while, and vanished. Mr Barrett rapped loudly on the table and rose to his feet. He asked all present to charge their glasses for the King's health, a 'loyal toast with musical honours', as the programme declared, and all rose in company to comply, whilst Squire Firth's cornet players blared out their fanfare. They drank and subsided. The murmur of renewed conversation had barely begun when Mr Barrett rose again to announce a toast on behalf of the Craven Agricultural Society and Farmers' Club and Officers to which the Reverend Morris beside him would respond. Cheers and deep drinking followed as the Reverend Morris, his stomach swelling above the table top, took a pulpit stance and began to speak.

He looked beamingly around at this scene of decent enjoyment, of good-hearted, good-stomached living. 'Well, well,' he said, and a chuckle of response ran around the room. 'Once,' he went on, 'once I was a young curate, just out of university.' He paused and nodded at them in mock solemnity. 'Never,' shouted a thick voice from the depths of the hall, and the Reverend Morris who had been rather waiting for this, shook a finger in mock reproof. 'I was indeed a young curate, a young, unmarried curate and assisting the rector in my very first parish. Quite some way from here, it was, far away indeed . . . . Now very early on I was given the task of making a preparatory examination of those about to get married, a simple matter, you might think, of factual investigation and record. At that time, you will understand, I had considerably less knowledge of the blessed state of matrimony than the young couples whom it was my duty to question. You must bear that in mind.' An anticipatory chuckle ran round the hall.

'One day a young man came to see me. "Well, well," I said. "Can I help you?" He was clearly from the poorer classes and his intended was a servant girl who could not get leave, for he came alone. I began to go through the prescribed questions, for it was a routine already, even to a very young curate, and in due course I asked him "Are you a bachelor or a widower?" He was obviously puzzled, for he shuffled his feet and scratched his head. I repeated the question. Then at length he replied. "Nay," he said extremely slowly, "I'm a warp-dresser. . . . That's what I am, a warp-dresser." Now I. . . . The Reverend Morris's further words were drowned by a wave of laughter.

'. . . Now I . . . now I,' he went on manfully, 'I am a clergyman, but that does not prevent me also from being a husband, nor does it prevent me from being a farmer, and in that capacity I appear before you today!' He paused for a mouthful of wine. 'In that capacity I respond to the toast we have just drunk to the Society and its Officers and I express the thanks of us all, all of us here to our sadly absent President, Lord Hothfield, and to my fellow Vice-President Mr Barrett for the hospitality we have been shown.'

John Tawney began to lose interest as the Reverend Morris went on to consider the state of Craven agriculture, and to doze, like many others in the hall, when John Bonny Dewhurst rose to answer the toast of 'Skipton, Town and Trade.' So much speechmaking on such a hot day was more than the brain could bear. It was not until the formidable voice of Walter Morrison began to speak on behalf of 'The Exhibitors' that he could raise sufficient interest to follow what was being said.

'I have found considerable difficulty, Mr Chairman, in reaching an understanding of the logic you have used in asking me to respond to this toast, and I have come to the conclusion . . .' Walter Morrison shuffled his bulk into a more comfortable stance. His eyes twinkled in anticipation. 'I conclude that it can be nothing else than the butter!' Curious grins were turned towards this giant, unconventional figure. 'The butter it must be, for this year, as you will be aware, I have provided a special prize for the best butter, having a personal and peculiar reason to bless our Craven food and air.

'Many of you are aware that I have suffered from an attack of tuberculosis, fortunately a mild one. I had no idea of the strange tortures that disease and medical science between them can impose upon a poor sufferer. Imagine to yourself my friends, the foul air cure . . . "the foul air cure", what a contradiction in words. To be closely confined in a windowless room, tight sealed, inhaling one's own exhausted air, to the accompanying stench of decaying flesh. That is an exceedingly unpleasant experience.

'Then to be offered the alternative of a diet consisting primarily of barley water, camomile tea and thin, unbuttered toast, that is no solace to a man of my disposition, who loves clean air and good food equally.' He paused to contemplate his own well-constructed stomach at this point and pat it complacently, giving a rumbling laugh and leading his audience to laugh with him. 'Now this explains the butter, and hence the exhibitors. The air of the hills and plenty of good food will do what closeness and starvation cannot do, and so I give my thanks in reward for rich yellow butter.'

XVIII

A DROWNED SPIDER

When Joady stirred again it was with the confused impression that some giant person was bending over him. He heard a voice calling his own name, a voice he knew, and opened his eyes. There, above his head, hung the underbelly of a great, black, heavy cloud, about to bust open. The voice called again. 'Joady . . . Joady Thacker?' It was his master's. 'Here,' he croaked, 'here I am.' Simon came towards him. 'Joady,' he said, 'where the hell have you been, you daft ha'porth. It's going to pour bucketfuls any minute. Come on, get inside!'

Almost as he spoke the first fat, heavy drops hit the sprawling cowman. Splat! and splat again, full in his face.

Joady rolled over and looked round. His stomach felt better, his head was clearer, but his legs trembled. Badger had vanished and as Joady surveyed the field he could see crowds of people gathered around the horse ring, but the threat of a downpour was beginning to break open the crowd. They were scattering hastily towards tents and awnings, umbrellas raised, coats hastily shrugged over still-sweating shoulders.

Too late. Down it came. The resolution of days of ever-increasing pressure and ominous heat. Down in a volley of heavy drops to be followed by a curtain of water that swept across the field. Water bounced from the dry baked earth and ran away in streams, turning the paths into a slippery, moving carpet. Everyone rushed for cover.

The rain beat down relentlessly, walling off tent doorways like a beaded screen. Heavy pools began to collect on sagging canvas roofs, growing deeper by the minute, pressing against poles, pulling taut all guy ropes, pregnant with threatened disaster. Common threats brought common action and the menfolk, eyeing these gathering lakes above their heads, began swaying the canvas to and fro until the water broke sideways over the edge and hurtled down to the ground amidst cries of 'there she goes!' And still it rained.

At first steam had hissed from the iron range of the cooking stall. Now it was still and cold, but Joady stood in the rain beside it. His clothes were totally soaked, clinging to him everywhere. His scanty hair was plastered to his skull. His scarecrow frame seemed to tremble under the pounding as he stood with hunched shoulders and stared out towards the invisible hills. But Joady was happy. The ache in head and heart, the hollow, resounding cavern, had vanished. He felt clean, washed free of guilt, grease and vomit. Slowly he gyrated in a circle under the falling water, spreading his arms, ignoring his master, who had been calling to him from the entrance to the beer tent, 'Come inside, you idiot!'

A shining line of white appeared behind the grey curtain of rain. The heavy black canopy of cloud at once seemed to be sliding past at a tremendous rate. Within seconds the rain drew away, the distant hills were visible, crowned in blue sky, and the sun shone down with an evening light. Steam rose from the ground, drops sparkled on every bent and broken blade of grass or summer flower.

Tentatively the crowded people edged out from their shelters, peering up at the sky in anxious search for any token of a new cloudburst. But the humped black tailings of the unexpected storm were already fast vanishing towards the east and the sky was clean, luminous and peaceful. They sighed, looking around at the muddy, debris-strewn desolation between tents, at their stained and steaming garments, the ruins of stalls and goods. Joady, his ecstasy vanished, lowered his arms, gradually spun to a halt and shivered in fresh sunlight as his damp clothes clung to him.

Slowly the ordinariness of an extraordinary day was re-established. Everyone began to drift towards the ring. The band slithered back up that perilous slope, their instruments held high, wiped clean once more, lovingly protected by soft cloths from any risk of further contamination. Children ran about splashing with their boots in the mud, the little boys taking care to soak their skirts as they splattered each other liberally. Mothers screamed in angry remonstrance. Fathers pretended ignorance. The first few notes of tuning, short blaring phrases from the brass, a muffled roll of drums, the odd tinging note of a triangle, rolled downhill from the band station.

It was not a scream, nor a shout, but more of a confused, buzzing announcement. Everyone stopped, suddenly attentive, and, again like ants, began to converge towards one particular spot at the fringe of the outermost line of stalls, close behind a ruined confectioner's booth. The trestles had slipped and fallen sideways under a mass of water that had gathered in the canopy and torn it down. In the mud lay melted jewels of many coloured boiled sweets, shining now in the sunlight, broken slabs from the mounds of pink and white coconut-ice.

Gran Bargh lay on her back beside these ruined treasures. Her mouth hung open, tongue lolling into the mud, and her eyes were staring into a puddle, a mere hoofmark full of water in which a demented beetle was gyrating upside down in iridescent circles. One hand was outstretched, fiercely grasping a sugary lump of coconut-ice with skeleton fingers, so fiercely, indeed, that the flaked and sticky sweetmeat was squeezed like paste from between her knuckles.

'She's fainted,' said someone, as onlookers gathered in a curious circle about the spot, 'who is she?' 'Nay, I reckon she's dead,' said another voice. A buzz of conversation followed. 'Poor thing . . . she's ugly enough . . . ugh, I don't want to see . . . Henry, take the children away . . . reckon it were t' heat . . . perhaps a thunderbolt . . . nay, that's an old wives' tale. . . .' Then a shrill lad's voice, breaking under pressure of excitement. 'Let me through . . . let me through. It's me Gran!' and Josh was out at the front, looking in fear, amazement and a deal of disgust at this sprawled, deflated figure. Reluctantly he touched her grubby cheek, drew his hand rapidly away and called in a small voice, 'Gran . . . Gran it's me.'

'She's dead, lad,' said a comforting voice, 'don't take on.' But Josh was frightened. The scene was so strange, the crowd of unknown faces pressing close around, staring, devouring with their eyes, the damp steamy atmosphere, his memories of the day, first the dream picnic, then the storm, and now this. It was like stumbling into a nightmare and he half believed they would all vanish, all the glassy faces, leaving his Gran to wake up, edge unsteadily to her feet, curse him sharply and make off.

'It's me Gran,' he repeated, and the sensible woman behind him responded quite naturally, 'Why, the poor old thing . . . and are all these folk just standing and staring. Come on, let's make her decent,' and she bent to pull the dishevelled clothing straight, to re-arrange the sprawling limbs in a semblance of sleep. 'Are none of you men going to get a doctor and help get her somewhere else?'

There had been something inhuman, unreal, about this corpse but now they felt confronted only by familiar death and so knew what to do. 'Josh lad, what's been laikin at?' Another voice, an arm laid on his shoulders, a cough, as Bert Ridding stood by his side. 'Oh, Mister Ridding, I'm right pleased to see thee . . . Me Gran's dead . . .' he pointed, half-ashamed, at the now tidy body where that helpful woman was just attempting to remove a mess of coconut-ice from the desperately clenched fingers, remembering that he had left his Gran to fend for herself whilst he went off with his new friends and had never even sought out Bert Ridding after leaving him at the gate.

'Hast sent word to thy Mam?' asked Bert. 'No,' replied Josh, shaking his head, 'no, I've only just come.' 'Run along then, lad. She'll not be that grieved, I reckon, but it's best she be told right away.' Bert gave him a friendly push on the shoulder, setting him free to run like a boy again through the slow-moving crowds, out through the gateway, down High Street and back to Roger's Yard.

XIX

A MAGNIFICENT DISPLAY

A shower of rain never did anyone much harm. Nodding their heads sagely to all they met, propounding such wise platitudes, the Reverend Morris and Mr Walter Morrison carried their own good humour into every corner of the field. Shepherding, cajoling, smiling broadly, they did their best, sending stewards panting off with urgent instructions to re-commence the programme. At length they summoned the bellman to ring a welcome to the new beginning, for the next item was to be the display and demonstration of farm machinery.

Then it was that Walter Morrison stopped suddenly in his peregrinations and frowned as deeply as if he had just foreseen the ending of the world. Other committee members also halted in the imminent danger of cannoning into him. 'The tractor! Samson Yates!' he exclaimed. 'I've seen no sign of them. Why aren't they here? We've been so busy with all this wretched business that I never thought to check.' 'Don't worry, Morrison,' said Mr Barrett, 'these machines are always temperamental, like the ladies. She'll be keeping you waiting at the altar until the last minute.'

The assembled committee members laughed at this sally, and even Walter Morrison himself smiled briefly from the depths of his despair. For Barrett was right, as always.

How Sam Yates had sweated in the midday heat, pausing only for a drink of lemonade, for he wouldn't touch beer when working on a machine; how he had eased and greased her joints, tightened up, tried her several times and at length, plying the blowlamp, got her to start; how he had raced across Skipton at a full three miles an hour as the thunderstorm broke above him, soaking him till the peak of his cap hung over his nose and his blue apron clung desperately to him – none of this would ever reach Mr Morrison's ears. But here came the new tractor, whirring into the showfield at the perfect moment, gaining the maximum attention as everyone flocked to see her.

'There she is, Walter!' 'Yes, I have eyes!' and he strode impatiently over to meet his engineer. 'Everything all right, Sam?' His despair and indignation had vanished as soon as the missing toy arrived. Sam was grinning. 'Yes sir, we were delayed at the canal.' That was all he would say of the troubles he had encountered. 'Well, we'd better show them all what she can do, eh?' 'That we will, sir!'

Sam drove the tractor up to the demonstration area at the far top corner of the field, beyond Storems Laithe, where the ground was nearly level. It could be seen from many parts of the field and Sam kept the engine running, not wishing to employ his blow-lamp in public. The piston and crank were vibrating with movement, the flywheel whizzing round. All this called out for attention and Sam stayed with his charge religiously.

He was a little disheartened by events, for the ground conditions had deteriorated and he feared the display might be a disaster rather than the triumph Mr Morrison expected. After all, he'd had no time for proper tests. The earth had been hard enough before. Now it was covered by a thin layer of slippery mud and vegetation, a most unpleasant combination when he was expected to demonstrate ploughing, mowing and load hauling.

The traction engines, which weighed three or four times more than his tractor, were struggling to mount the rise from the back entrance to the display area. A handful of navvies marched alongside them, shovels over their shoulders. They had been given free entry for the afternoon in return for help with this part of the proceedings. Sam knew the machines.

Both were Fowler sixteens with compound engines and worm and wheel steerage; good designs. They had steam well up and whistles blowing to clear the way. Sam tipped his cap to the drivers as they finally reached him, leaving an enormous pair of grooves behind them. The second engine was towing a six furrow balance-plough.

The navvies stationed themselves behind, leant their shovels against the wall and began to pass a substantial, bulbous bottle from hand to hand. 'This is no time for digging, no time at all,' said Tony Corrigan as he wiped the wheel spray from his cheeks. 'No rock ter shift, any road,' grunted Little Charlie from between his enormous shoulders. They settled back against the wall, Bacon Billy chewing as always.

The demonstration areas had yet to be marked out and the steam ploughing rig set up, which would take half an hour at least. 'You'd better get things moving,' said the first steward to arrive, 'the committee are on their way.' Mr Edgar O. Hudson, chief engineer of the Yorkshire Dales Railway, was with him, having been made an honorary steward for this occasion, and he took charge.

'Corrigan,' he said, 'get your gang up.' 'Yes, sir, certainly sir, what do you want us to be doing?' 'I'll tell you.' Mr Hudson was normally a man of few words and quick commands, so Corrigan waited. 'Take those stakes and pace out over there. I want a hundred yard by hundred yard square marking, then we'll split it up.'

Whilst the display area was being marked out the second traction engine moved to the far side of the square and stationed the two-winged balance-plough, with its man-high wheels and double seats for the driver, ready to begin the first furrow. The two engines were opposite each other. Now the wire ropes from the powered cable-drums under each of their bodies would be connected to the plough and the system was ready for operation. The huge plough would cut six furrows at once. It was a double plough, with one set of shares facing the opposite direction from the other. Pulled towards the first engine, it cut its six furrows, then the driver changed seats, the balance was carefully altered and the second engine began pulling, so drawing the reverse set of shares through the earth.

After each double cut the engines must both move forward to uncut ground. It was a cumbersome, elaborate process, requiring two engines and three men, and not really suited to rough land. Better in the vale of York than the dales. Hence Walter Morrison's encouragement for the new light tractor.

The committee had arrived by now and were inspecting the new machine. 'Now you will see the contrast,' Walter Morrison was telling them. 'This is much simpler, much more simple. Fowlers assure me that it will pull a three-furrow plough at two miles an hour. One engine and one man only. A splendid device,' He slapped the cylinder box. 'All ready, Samson?' 'Yes, sir, but the ground's tricky, sir.' 'You'll manage. Never fear.'

The navvies were back now by the wall, in attendance should any digging out be needed, and larking about. 'What's this,' said Walter Morrison, 'what have we here?' Tony Corrigan stood up. 'Come as ye told us, that we have, sir.' 'Good. Have you brought your shovels?' 'Surely, sir. Though the earth, she's hard and muddy both.' 'No worse than the railway works at Embsay is it? But here you are.' A half-crown changed hands. Tony Corrigan saluted.

'What do you think of the rock at Embsay now, Hudson?' said Walter Morrison to his engineer, who had just appeared. 'Nearly all cut through,' was the cautious reply. 'Couldn't have foreseen it in survey. We'll catch up soon enough.' 'Yes . . . yes, yes, we'll have that railway working any time, Hudson.' 'That we will, sir.' 'And a real benefit to Craven!' 'Certain to be, sir.'

Another spectator, unnoticed in his insignificance by the great prime-mover of all this progress, had wormed his way into the very heart of the business and was studying the machinery with total fascination. Josh was back at the Show. He had told his Mam of Gran Bargh's fate, observed no reaction to the news he had carried at a breathless run, slipped away again, evaded the careless stewards and was right at the heart of events once more.

The world felt a different place to him. The whole day, new friends, new interests, the heat, his Gran's death, all seemed to have fermented inside him and he could have sworn he was five years older than the lad who explored the depths of Eller Beck, so long ago it seemed. He was almost hypnotized by the moving elements of those great machines. Josh felt an urge to emulate the quiet, competent, self-assured men who tended them.

He crept up beside the driver of the red and green tractor who was bent over a shackle behind his engine. Samson Yates looked round. 'Hey, you lad, pass that hammer.' Josh dived eagerly at the tool and handed it over, but the linkage was just too far out for Sam to drive the pin through. 'Can you hold this?' he asked Josh. 'Aye, sir. Course I can, sir.' He was a small lad, but grim determination shone in his face. 'Right.' Sam swung up on to the driving stand, let off the brake until the tractor rolled back half an inch, put the brake back on and jumped down. The holes matched. Josh was keeping them fiercely together. Sam slid the pin in, then smiled at Josh's clenched jaw. 'You can let go now,' he said. 'What are you going to do, sir? What kind of engine is it? Does it plough by itself?' Sam took his shoulder. 'Come on, lad. That's a load o' questions. Dost' want a course in engineering?' 'Yes please, sir. Could I, sir? Can I learn?'

Walter Morrison had heard. It was perhaps the luckiest moment in Josh's life. 'What's this? What's this? Who's this young man?' He asked. 'Don't know, sir, he's been giving me a hand.' Walter Morrison surveyed Josh's ragged clothes and eager face with kindly interest, saying, 'Who are you, my fine fellow?'

'They call me Josh Bargh, sir.' Then, unable to repress his keen excitement, he added. 'Is that your engine?' 'Yes, indeed,' answered Walter Morrison. 'It's wonderful,' said Josh. 'You like it, eh? It can do all kinds of things.' Walter Morrison was speaking as one enthusiast to another. To the outside observer it might have seemed comic, this conversation between the rich middle-aged man, former M.P., and the small ragged boy, but those who believe see the world transformed.

'Where do you come from?' 'Roger's Yard, sir.' Josh rather mumbled the words. 'Aha!' Walter Morrison was not to be frightened off by such beginnings. He was a man who believed honestly in self-improvement and that no one was beyond hope simply because of his origins. Certainly a boy could not be condemned for his birthplace. 'What do you intend to do when you grow up?' 'Don't know, sir. Be a half-timer right away. Happen go to t' mill . . . Hope not!' 'Why not? It's a good job at the mill, isn't it?' 'Aye sir, but . . . I reckon there are other things.' Josh looked wistfully at the tractor and Samson standing at the controls.

Walter Morrison smiled to himself. He had seen that look. 'Would you like to ride with Samson here?' he asked. 'Oh yes, sir. Could I, sir, please!' 'Very well, Josh Bargh, you can get up and help.' Such a thankful smile Walter Morrison swore he would remember all his life. Josh swarmed up the rear of the tractor to stand in glory beside Samson Yates. Walter Morrison took out his pocket book and noted the lad's name. He frequently suffered from people taking advantage of his good nature and so he added a question mark to the note.

'Please, sir, who was that?' Josh asked as the tractor began its first run. Sam struck his forehead with the flat of his hand. 'That? Why, that were Mr Morrison, the millionaire! He's my boss.' 'Were it really?' 'True enough. Now then, you watch out, young feller, and help. I can use thee. What's thy name?'

So the display went on. Certainly the tractor stuck now and then, but so did the traction engines, and it ploughed well over half the area that the balance plough covered, and with the labour of only one man and a boy. At length Samson drove his tractor round in a victory lap of the demonstration area whilst Josh stood beside him, serious, responsible and proud. They drew up before Walter Morrison, who was obviously delighted with the results.

Tony Corrigan and his mates had already begun to drink Mr Morrison's half-crown. Now was the time to repay this kindness. Tony set them to cheering, a hoarse and continuous noise that was caught up by the spectators and repeated round the enclosure. Walter Morrison beamed with happiness at such spontaneous recognition and raised his hat high in the air.

Sam climbed down, leaving the engine running. 'Went very well, Samson. Very well indeed. How did you find her?'

'Quite light to handle, sir. She pulls away easy. I don't know how she'd manage on really steep ground.' 'Hm . . . perhaps we'd better do some tests out at Malham.'

The Walter Morrison turned to Josh. 'Well, young man, what did you make of it all?' 'It were right grand fun, sir. An'. . . I'd like to know more about them engines.' Sam interjected, 'He's been pestering me with questions all the way round, sir.' 'Keen are you?' Walter Morrison asked. 'Yes, sir! I am that, sir!' 'I see you are, and I like to see it. If you come to see me at Malham, . . . anyone will direct you, we can perhaps discuss it with Samson here. You never know, we might make an engineer of you. That would be something in Roger's Yard!'

XX

BUTTER

The Stockdales and everyone else were on their way to the show ring for the final parade, the selection of the grand champion from among all the animal entries and the award of prizes. It was later than usual at the Show, but the storm and then the display of machinery had delayed proceedings. 'I wonder who's won prize for t' butter,' said Martha thoughtfully, for no coloured cards had been seen in that part of the tent. Walter Morrison was keeping a little excitement to the end. 'I bet it's that sour-faced woman we met.' 'Nay, wi' a face like that she'll turn milk off right away, make nowt but whangy cheese.' Simon laughed at his own joke. 'You never can tell,' Martha replied.

At that moment the crowd suddenly parted as a lantern-jawed figure stalked towards them, black and stilted in his walk, forked stick held forward in anger. 'I thought it were all agreed. We'd shaken hands on it, Simon Stockdale,' the harsh voice grated. Behind him ambled the innocent cause of one more disturbance, mildly chewing the cud, wafting his tail against the flies. People drew back hastily from the bull who had terrorized the whole of Skipton that very morning.

'Mr Stockdale,' this was a thin, shy, but determined answer to the black-jawed cattle-dealer and his claim. Joady Thacker held tightly on to the halter round Jasper's neck. 'Mr Stockdale . . . tha's not sold Jasper to Jemmy Rattle?' 'I've bought him,' expostulated Jemmy. 'He's mine . . . agreed and paid for, shaken hands on it!' Jemmy heaved on the halter, raised his forked stick and glared blackly at Joady Thacker who stubbornly stood his ground, bleating, 'Mr Stockdale . . . Mr Stockdale.'

'Let him go, Joady! Didn't I tell thee we'd sold him?' Joady's face fell, somehow he felt he had wakened into a new, harsher world, where old friendships no longer gave security. His mouth dropped open for a moment. 'So tha has sold him, then,' he said, 'I didn't believe it,' and his hand dropped the halter. 'How much did he offer thee?' 'It's no business of thine, but . . . seventeen pound.' 'Th'art witherin him away . . . just witherin him away like . . . like putting a good coppy on t' fire. All burned up for nowt.' 'Shut your mouth, Joady! Stop clackin'. It's all shaken on, and this is neither time nor place . . .' 'Aye . . . I can see that. Reckon it were my fault . . . I should've kept wi' him. All day, right from t' start. Reckon I'm the Judas today.' Joady beat his forehead with the heel of his hand, turned about and vanished through the crowd. 'Hey, where are t' going,' shouted Simon after him, but he paid no heed.

'Right, Simon Stockdale. Is all t' fuss over then? Wills't keep thy bargain? The harsh voice spoke in his ear. 'Aye,' he replied softly, absentmindedly, for he was attempting to fathom Joady's action, 'aye get you gone! But mind, he must be in t' ring for t' parade or I won't even get second prize money.'

Before Jasper could be taken, Sally ran over to him and put an arm round his neck. His blue-black tongue curled against her arm as he reached out for the daisy chain she wore. 'Don't eat it . . . no!' she said and then, as a farewell offering, took the chain over her head to give it to Jasper. It would not even span his horns but hung uncomfortably on one side and trailed across his eye like a swarm of flies. He shook his head and snorted as he was led away.

'Dost' think we got a decent price for Jasper?' Simon was seeking Martha's support. 'I don't know. It doesn't seem much, not for all that bull, but then, it never does. You should know.' She paused in thought. 'I wonder what's got into

Joady.' 'Well, he's been daft all day, not himself at all.' 'I can see that,' Martha replied. 'I hope he's not going soft in the head.'

The parade was already in progress when they reached the show ring. Simon was waiting for the cattle, mostly local farmers' shorthorns, matched pairs of bullock stirks, milking cows, and a couple of old Craven longhorns that proudly led the display. At length the winner of the class for the best shorthorn bull under eighteen months was announced and the champion lumbered complacently round the ring, his belly swinging, little eyes maliciously watching the circle of faces round him, like a prizefighter before the first round. Jasper stood in the line, held by one of Jemmy Rattle's men, and Simon went over to collect the prize money, a nominal pound. He was quiet when he returned.

Martha, wedged between a billowing farmer's wife, a distant acquaintance, and Simon's muscular side, with Sally perched unsteadily on a half-broken wooden box before her, sighed heavily. Simon's face had taken on a stony, guarded look. His professional eye was inspecting the Storiths bull and he speculated on the probability that Jasper would have won if he had come in proper shape.

Sally began to cry quietly at first, then with her whole body shaking. Martha became aware of the sobbing at her side. 'Why lass, what's got into thee? Are t' fretting over Jasper?' Sally's answer was so soft, so demanding of love, as to be almost inaudible. '. . . I don't know . . . I just feel sad . . .'

Martha's grey head was bent close over the golden hair and she was smiling gravely to herself, deep in the pleasure of giving comfort and sympathy. Meanwhile, the procession had disappeared and the roll-call of other prizewinners began, a periodic series of bellows from the crier as he announced the victors, punctuating the background gossip and chatter of the spectators. Something in this external noise made Martha raise her head in surprise. Simon was staring at her with an amazed, wrinkled mask of surprise, his eyes shining.

There it came again, cracked from the bellman's parched throat and leathery mouth. 'Special prize . . . best butter,' the vowel sound deepened as if he were describing some kind of footgear, 'Mrs Stockdale, High Dubb Farm, Littondale.' Martha's stolidity vanished. She jumped up, waving, shouting. 'Here . . . Here!' 'Aye . . . it's t' butter, lass,' Simon confirmed as she grabbed him round the neck and hugged him tight. The stout neighbour guffawed as she watched them and took Martha by the shoulders, saying 'Come on lass, get thy prize . . . they'll not wait all day.'

She was under the rail, free of the crowd, dragging Simon and Sally by either hand, then exposed in public, out in the sunshine of the ring. A halo of shyness ringed her homely figure as she straightened her back and braced herself to stalk at last, stilted, upright, red-faced and red-necked across the echoing plain. Every simple housewife among the crowd felt the meaning of those instinctive gestures, and their sympathy went with her like a wave to the very edge of the rostrum.

The sea of faces blurred in Martha's eyes. Her world narrowed into a pair of polished, mud-splattered boots and the baggy trunks of a pair of stout-legged trousers. These she saw with intense clarity, down to the individual specks of pepper and salt material.

Slowly she raised her glance over the spreading waistcoat, the heavy gold watch chain linked to its hidden hunter, the floppy collars, shaggy beard of brindled whiskers, to the face of Walter Morrison. 'Mrs Stockdale,' she heard the bulky voice roaring in her ears, 'I'm delighted to meet you. I consider it an honour to have a neighbour with such talents. Your butter is delicious, truly delicious. . . . I hope you and your family eat plenty of it yourselves. . . .' 'We do that, sir,' answered Martha, smiling still. 'Did you have the doubtful benefit of hearing my remarks over lunch? . . . No?' as Martha shook her head. 'Ah well . . . I'm extremely pleased to be able to make this presentation.' He held before Martha's eyes, taut between his fingers so that it was spread large as a tablecloth, a banknote of the Craven Bank, a promise to pay the bearer five pounds, in elaborate copperplate on a crisp sheet of paper, with its proudly emblazoned symbol of the Craven Heifer. 'I acquired one of these on purpose,' said Walter Morrison, 'with the intention of celebrating butter equally with famous cattle. You wouldn't like to sell me this batch?' He swung

Martha's basket in his hand, 'I'd be glad to take it home.'

Martha paused, considered, wrinkled her nose and pursed her lips. Then her eyes began to twinkle as she spoke. 'I don't reckon I will, sir,' she replied. She paused for a moment before continuing, 'But I'll tell you what . . . I'll give it thee with my blessing.' Walter Morrison, having looked slightly surprised at the rebuff, now threw back his head and roared with laughter, clasping his watch chain as his stomach swelled with mirth. 'Thank you, Madam, thank you,' he shouted and held up the basket for all to see. Cheers and applause followed from the ring of fellow worshippers of good-living that surrounded him. Martha walked steadfastly back to her place, the five pound note flapping in her hand.

XXI

CELEBRATIONS

'Nay, lass,' said Simon as soon as they had reached the safety of the ringside, 'I'm right taken with thee,' and he kissed her cheek. Sally was hugging her closely round the waist. This was a day of glory after all. Tears had vanished. Martha, as her elation subsided a little, was meticulously folding the celebrated Craven Heifer and stowing it away in her bosom. 'It'll stay hidden,' she said, patting the place, 'till it's needed.'

'What are you going to buy with it, Grandma?' 'Nay, I'll ponder a while . . .' Simon grunted in amused comment. 'What's wi' thee?' 'Where's yon sovereign Aunt Bertha gave thee on our wedding day?' Martha eyed him sharply. 'It's at home . . . tha knows it well.' 'Aye, and will this one be kept for a rainy day wi' the others?' 'Aye, just so!' 'Were it not a rainy day that year we had to slaughter half the cattle and were at our wits' end to keep on?' 'Nay!' interposed Martha staunchly, 'don't talk of that . . . I shall know it's truly a rainy day when that money's truly needed.' She nodded up and down once or twice to emphasise her determination. Simon rubbed his hands in admiration. 'In the Lord's name . . . what a woman,' he sighed as he gave her his arm. 'Come on. Show's nearly over. Let's take a walk down High Street.'

As the day cooled and the beginnings of an evening chill blew across bare arms and necks still flushed with the heat of the day, people were leaving fast. The sky above the distant hills was slowly thickening into deep blue obscurity. Horses flicked their tails restlessly at the first swirls of thinly whining midges. Daytime flies were buzzing sleepily in search of a night's refuge.

The Stockdales ambled out of the gate and down the Bailey, receiving now and then the congratulations of acquaintances on their success. Sally, skipping at every other step, had buried her sadness over Jasper and became a happy granddaughter again. Then some obstacle on the pavement brought them to a halt. There was an eddy in the slow-moving crowd as folk broke and passed each way in front of them. Slowly a space cleared until they stood right above the cause of this interruption.

Two lads crouched at their feet, a wooden hutch between them. Bertie Whinsby was holding close against his chest a large, sleek rabbit. His head was bent protectively over it. One of its long ears hung down beside his cheek. It was torn and bleeding. He was blubbing loudly. His fair-haired companion rapped a knuckle on his forehead, saying contemptuously, 'Stop yer yarmin . . . stop it, Whinsby.' 'Just look at her,' came the tear-filled reply, 'there's gurt holes in her ear. She's still trembling. I can feel it!' 'It'll mend,' said Josh, 'you'll get over it. . . . Any road, th'art trembling thiself. After all, look at me. I'm not trembling. Me gran died today and I'm not blubberin' neither.' The bigger lad gently stroked his pet. They both seemed calmer. 'Nay,' he replied at last, 'nay, but that's different. A grandma's not like a rabbit. Tha hadn't brought her on specially for t' Show.'

Taken aback by this remark, Josh sat slowly back on his haunches and whistled softly to relieve his feelings. Then he stroked the rabbit himself and laughed. He looked up and noticed for the first time the Stockdales watching with amused expressions the scene before them. He stared, owlish, then grinned. 'Sally, Mrs Stockdale,' he said, 'I were looking for thee.'

Josh blinked up at them. 'I don't want to go back home!' he said, rubbing his chin desperately with a clenched fist as the grin faded and he gave thought to past events. 'She's there, she is. They've shoved her under t' stairs, but her feet are sticking out . . . I'm not sleeping up above . . . she'll be staring through t' floor boards . . . I know her . . . she'll be watching.'

A feeling of sympathy for this strange lad they had met only a few hours ago left all three in silence. Then Simon put a consoling hand on his shoulder. 'Have no fear, lad. The dead won't rise in Skipton, and . . . ye've had a hard day . . .' Josh looked stoically from him to Sally.

'And me mam's drinking . . . She'll end up just like her . . . I can see it.' His voice was breaking slightly, jerking between squeaks and hoarseness, under the pressure of anxiety, as if a vision of the black future, a continuing threat of scruffy poverty, of drunkenness, weakness of will, and an ending like his grandmother's weighed him down.

'Come on, lad,' said Martha, 'we've got something to celebrate. I'll tell thee all about it and then tha can help us enjoy ourselves this night . . . till we have to go home, any road. Come on, let's walk on.' So, Bertie and his rabbit quite forgotten in the dust, Josh joined the Stockdales in a solemn parade down the Bailey-walk. Quite naturally he found himself beside Sally, with Simon and Martha on either side. It was rare for him to be treated as a child, member of a family and naturally sheltered, and so this simple, automatic arrangement at once added a warmth to his heart and a majesty to his pace as he attempted to match Simon's long stride. Sally winked at him and his heart pounded in sudden delight.

'Now hist,' said Martha, 'didst' know I won first prize at t' Show for best butter?' 'No,' said Josh, then added in compliment, 'I'll bet it's grand butter.' Martha chuckled. 'Not bad, tha must taste it one day.' 'I'd like that.' 'Aye,' she replied, 'but guess how much the prize was?' Martha enjoyed recounting her triumph. 'Nay, how can I tell?' asked Josh, then, sagely conjuring up in his mind an enormous sum, round gold coins with the Queen's head, big as spectacle lenses before each eye, he added, 'two guineas?' 'More nor that,' said Martha with a touch of scorn, 'it were five pounds in a banknote. I've got it here. . . .' Josh whistled. Martha carried on, 'Summat to celebrate, eh, lasses and lads. We'll have a bite to eat and go on to t' fair. What dost' a say to that?'

'Oh please . . . yes, Grandma . . . can we?' Sally held Martha's arm and now she took hold of Josh's. 'Can we go, Grandad?' She knew there might well be some moral disapproval to be overcome and Simon had been known to put his foot down at times when womenfolk let their imagination carry them too far. Sally smiled particularly wistfully at her grandfather, her hair dragging softly over one side-turned cheek. 'Aye . . . well,' responded Simon, in difficulty, as he shook his head to re-order his thoughts. 'Why not?' He was answering his own objections. 'Mebbe t' fair's not a holy place to visit, but then, there's no evil unless the heart's touched by it.' The moral choie had to be taken with due seriousness, but Sally's wistful smile was still before his eyes. 'Nay , it'll do us no harm!'

The sudden kiss took him by surprise. Silky hair caught on his stubbly chin and Sally giggled. He made an astonished face at her, saying, 'So thou would go among the tents of the ungodly, young woman?' 'Please, Grandad!' 'Well, perhaps we should get a bite to eat whilst I think it over. I've a mind to be tempted a little more. . . . Where shall we go for dinner, young man?' He spoke in the same joking tone to Josh.

Here was a chance to use his expert knowledge and repay a little of the unearned kindness. 'I know where we can go,' he said with instant inspiration. 'I reckon you'll like it,' he added confidently, and set off, dragging Sally's hand, towards the celebrated Globe Coffee Tavern.

The windows were filled with gaslight, glowing noticeably in the beginnings of grey evening light, welcoming as a fireside party in its mingling of tidy polished chairs and white tablecloths with groups of happy talkative visitors. 'This looks nice,' said Martha approvingly as she glanced round. The manageress, all graciousness to a family party, bustled over to escort them to a table. Then she recognized Josh, paused momentarily to register his previous incarnation and, unable to suppress her curiosity, addressed a question to him. 'Well, young man, are you treating some more folk with buried treasure?' she asked in mock severity. Josh grinned. He was happy to have brought his friends to such a gleaming palace and to be recognized as a regular customer. He wriggled in satisfaction on his seat as the menu was offered and discussed and the realization that he was truly promised another grand meal made him tingle from head to toe.

'You've been here before then, Josh?' asked Sally in surprise, and with a touch of respect for this town sophisticate. Josh told them the tale of the celebrated sixpence, earning a laugh from them all and a hearty slap on the shoulder from Simon. 'What a lad th'art,' he said, 'I mind now whan I was a lad myself. . . .' Then, realizing how hungry he had become after the afternoon's emotions and excitement, he set to work, digging a way into his food as if it were a vegetable patch, and the others followed suit.

As they dined they were also entertained, for it was the custom at the Coffee Taverns to have a reader employed, in return for his dinner and a few pence, to declaim selected passages from the newspapers and suitably edifying books to the assembled customers. The reader that evening was a feverish-looking man who dramatized everything that he read. Half his forefinger was missing and Sally couldn't help staring at his hand in fascination. The manageress told them in an undertone that he had once been a schoolmaster, an educated man, but had lost his post through drink. This added spice to everything he said. They sat and listened for a while, enjoying a leisurely cup of tea after their pudding.

Simon paid up for the feast, since the Craven bank-note was to be kept in reserve. Then they sailed out into the thickening darkness of a warm late-summer evening, feeling comfortable inside and out, filled with pudding and a peaceful apprehension of the fair to come. In High Street they stood for a moment of quiet preparation.

XXII

AT THE FAIR

The canopy of sky soared high above Skipton, arched with faintest lines of colour across the western horizon, darkening blue at the zenith with the slightest glimmer of stars. Black, lumpy outlines of the houses were half visible above the warm glow of framed shop-windows, the gaslight simmering yellow across dry pavements and echoed in the pools of light falling from regular lines of tall lamps set into the market cobbles by the roadway. Like fountains in a formal garden the colours splashed cold green on to the outer branches of the young lime trees as they shifted infinitesimally with faint movements of the cooling, night air.

Simon led his party forward once their eyes had become accustomed to the darkness and then edged through the line of horses and vehicles still creaking down High Street after the ending of the Show. 'What've you done wi t' cart, Simon?' asked his wife. 'I got Joady to take it to t' Black Horse stables once I'd found him.' 'Was he all right, Grandad?' Sally asked after her friend. 'Well . . . perhaps he was a bit shaken up wi' losing Jasper . . . or else mebbe he's sickening . . . any road, he was white as skim milk at the time.' 'What's he doing now?' 'Nay, how should I know?' Simon added, slightly nettled by this cross-examination. 'He's to meet us beside t' cart in good time to get home . . . tha can ask him thiself!'

They were nearing the fair. A swollen bubble of light hung

in the sky above them, glaring yellow behind the cut outlines of Town Hall and Red Lion. A reflected glow lit up the bastions of the Castle beyond, turning it into the backcloth to a stage, a flat-seeming, painted symbol to close the view. Yet this strange, magic light rose from a jumbled base of stable yard, sheds and dusty closes and a core of echoing noise that spilled outwards to summon attendance. Discordant rhythms, organ blasts, shouts, shrieks of mock terror and bellows of beasts were compounded into one continuous blaring note that made the heart beat faster to hear it, a challenge, invitation and welcome as from a sorcerer's cave.

A dark, narrow passageway led into the fair between two high, stark walls. Sally clung to her grandfather as they were pressed inside. After a pace or two they came upon the first, feeble glow of oil-lamps from the poorer stalls that lined the way with a light that served to cast sinister shadows on to the faces of passers-by.

The passage grew more dense with crowded blackness and Sally clung more tightly. Then a burst of white light revealed the true nature of these shadowy, suspicious figures. There shone a group of sturdy Craven folk, laughing and pointing before a pair of hissing naphtha lamps. There in a pinafore stood a little boy of six, shrieking his wares to attract the crowds and jingling a huge money sack of faded red plush almost as long as himself, resembling in combination of white dress and red a choirboy gathering in the collection. His shrill calls gradually became more distinct. 'Walk up . . .walk up . . . see the murders!' The murders? Sally cocked her head in surprise. Yes, the shrill, childish treble with its knowing undertones rang out again. 'See the terrible murder at the Red Lion. . . . 'Ere's Maria Martin lyin' in 'er blood. . . . Walk up, walk up.'

A stout lady with mottled, bare arms had just paid her penny and was bending over to look through the peephole at the tiny, illuminated stage within. 'Ooh,' she screeched, drawing back for a moment, 'it's right real. . . .' The little imp grinned to himself and held out the red bag for more pennies.

The way became another dark, constricted throat as they pushed past the light, a trenchfull of strange faces, murmurs and shrieking laughter. Oil-lamps gave half-formed glimpses of staring crazy show-people in their dirty finery; and a man with rats running over head and shoulders, bulging eyeballs and a red head. What could he be? Next came a thin, tired-looking young man who played the flute to a dancing bear cub. His begging-hat held only a few halfpence.

They reached a row of huge, white teeth hanging on a string, swaying to and fro, reflected in faceted patterns by the dark glass bottles arrayed on the stall below them. A porky, jacketed man with beads of sweat shining in the rolls of flesh that hung over the back of his collar, was bending down to look into a gaping mouth held up towards him like a fledgling bird's. A thick hand, the back raddled with purple stains, held the straining jaw of his patient, while the dentist's waistcoat hung baggily down, without concealing the swaying belly. His fat voice was full of hoarse confidence. 'Your masticators will be no problem, Madam, none at all. If you wish . . . pray hold still . . . I can make available some superbly manufactured dental replacements. You would not believe the relief . . . no, Madam, the pain will be transitory . . . pray hold still. In the meantime we must do what we can.' The face in his hands puckered and shook. Tears appeared in the captive's eyes. 'A simple extraction. Merely a simple extraction for the time being. . . . Do you have a shilling?' As the struggling prisoner reached desperately into her purse, the Stockdales and Josh passed on to the fair.

Jerry Croft, the fairground, unrecognizable as a world of magic, noise, colour, light and movement, swallowed each humdrum, cautious individual into its transformations. Sally was fascinated, absorbed by this powerful life of rhythm, sound and flaring brightness. Held imprisoned in this musical paradise – propelled more deeply as an organ, brass and deep red paintwork gleaming, eccentric cams rotating, belts flapping, worked the pleasure outwards to her senses – she swayed and wondered. The machine gasped shrilly. Sunburst patterns in striking yellow, framed by rings of painted studs, jerked into vision before blurring into motion again. Sally blinked and looked around her.

'Look Grandma . . . look Josh . . . just look at that!'

Hurtling towards them to the accompaniment of screams and raucous laughter they saw a solid, barn-door sized, representation of the Stars and Stripes seemingly about to crash down on their very heads like a toppling tree. They leapt back fearfully. High in the air, poised vertically in defiance of gravity, the huge, painted flag hung for a moment then, slowly, very slowly it began to swing away from them.

Exchanging looks of wonder, as eyes and brain attempted to make sense of this apparition, they followed its backwards progress, observing closely what it had become: a giant, canopied gondola, packed tight with people who were caged inside by rails as in a ship, suspended by iron rods from an iron-braced gantry. The ship was gorgeously painted in lemon yellow and seawater green with formalized roses in primary colours giving it an air half nautical, half canal barge. As they drew nearer, the Stockdales could see a sign above the gantry, shining in incandescent light: 'The Steam Yacht Columbia', it said, in golden lettering as swirling as the sea.

There, indeed, was the yacht, gimballed in iron, and loaded still with hilarious passengers, but with no sign of that first, astounding American flag. Its zenith reached, the yacht lurched back towards them, and as they drew to one side all mysteries were revealed. There she was, painted flat on the yacht's broad bottom a flag in red, white and blue, soaring once more towards the heavens to hover above their delighted faces.

'Let's have a ride . . . let's go on . . . Grandad . . . come on!' Simon, having passed through initial doubt and caution in this unreal yet very plausible world, the very replica of sin, had now reached a state of happy acceptance which would most probably be a cause of anxious self-examination as he walked the fells next day. In unselfconscious glee he gave an arm to Sally and climbed the steps to wait. The hurtling yacht subsided into calm. Ecstatic passengers disembarked on to its wooden quay. Soon the Stockdales and Josh were on board, whirling in state through a sky of streaming rainbow lights and cascades of blurred darkness. Simon chuckled to himself. Martha grunted with amusement. Josh and Sally, at first screeching their excitement, then silent, sat with shining eyes and a shared intensity of delight, hand in hand among the golden balustrades.

At last an exhausted farewell hiss of steam accompanied their return from the skies to the trampled grass of Jerry Croft. They wandered in dazed emptiness through noisy cries, crowded alleyways, floods of glaring brightness and shadowy corners. It was another summons, a trumpet note close at hand, that shook them back to life. Once again the sound came, close above them. It was the elephant. Martha laughed outright in pleasurable affection as she saw the comfortable form, the wise, kindly eyes and wrinkled skin. The beast, in reply, opened her rose-red mouth once more, curled high her trunk and trumpeted like laughter.

Behind her curved a circle of wagons, wheels as high as a man, close drawn, shaft to tailboard and forming the outer-works to a circular tent of painted canvas. On the wagons' sides hung more pictures, hunting scenes, where black panthers tore at each other with gory, snarling jaws against a background of rocky desolation. One sign proclaimed, 'Wombwell's Celebrated English and American Menagerie', for all to see.

Close beside the elephant sat a fine lady, displaying three feathers in a jewelled turban, a good deal of not-so-white bosom and a sign that read 'Ladies and Gentlemen 6d. Working people, servants and children 3d.' 'Shall we go in?' asked Simon. Sally and Josh both nodded violently and Martha squeezed his arm. Simon paused, as always, before committing himself, and at that moment a stately gentleman, in leopard-skin cap, emerged from the tent and began a stentorian oration. 'Walk up . . . walk up, ladies and gentlemen. This way to the greatest show on earth. See! See the great performing elephant! Children may safely ride on her back. See the lion and tiger both in one cage . . .' Pause for effect here. 'True! True or I'll forfeit a thousand guineas . . . Walk up . . . Walk up. Don't be deceived. The elephant is here. Walk up!'

'Which price am I to pay?' said Simon to Martha loudly. 'Are thee and me working-people or gentlefolk today?' 'Simon! Hist!' whispered Martha in shocked surprise at this light treatment of a serious subject. But she answered him properly, though in a low voice. 'Don't be mean, lad. We're celebrating, aren't we? Pay up sixpence apiece and forget it.' 'Right,' said Simon,' and I'll pay for t' lad. That'll be what I owe him for minding t' cart.' So they passed by the fissured legs of the elephant, giving her a friendly pat as they went, paid their fee and prepared to gasp at the marvels inside, as an aroma of rank flesh, hay, pungent stables and new paint welcomed them through darkness into light.

After more than an hour of staring amusement they emerged, chattering but weary, with wrinkled noses and flushed faces, to amble dazedly in search of fresh delights. 'Watch it, mate!' A burly fellow in stained blue sweater, spiky haired and pig-eyed, had barged into Simon. 'Watch where y'er goin!' A bulging, aggressive face glared closely at him and he was shrouded in beery breath. Simon's chin bristled. He raised his arms, fists clenched, and, looking coolly at this intruder, answered 'Say that again, lad!'

The burly aggressor hesitated in the face of such a sturdy figure and, as he did so, the woman on his arm pressed a swelling breast against his side. 'Come on, Badge,' she urged, 'come away . . . don't take no heed. If tha starts laikin an' thungin then I'm going home!' Badger was distracted by this threat to his newly recovered love. 'Come on,' she repeated as she pinched his heavy buttocks with all her fingers. Badger giggled, happy to turn aside from this staunch opponent, bending to whisper with his sour tongue into her fleshy ear. They drew away, leaving Simon, ignored in his anger, about to say something to more purpose.

The Stockdales were quiet for a few minutes. It had been a long day. Then, 'Can I take Josh on the roundabout?' asked Sally, as she saw the galloping horses rise and fall across a temporary clearing in the crowd. For her that day had become an intoxicating compound of light and dark, dazzling, threatening, but never less than intensely demanding of her love and attention.

They mounted the horses and circled the earth together in a shimmer of light and laughter. Glimpses shot in and out of mind: a carved, dog-like lion snarling in white fangs and red paint; a deep canyon full of sinister moving shapes; a flash of tinselled clothes and dazzling brightness as they rounded the dark side into full light. Then, at last, they saw and passed their own small island of security in a sea of strangers, the wise faces of Simon and Martha watching over them from the grass.

'What do you make o' yon lad?' asked Simon. 'Nay,' answered Martha without hesitation, 'there's no malice in him. I reckon he's a good enough lad, with some spirit to him.' 'Aye,' replied Simon, 'I've quite taken to him.' He chuckled. 'He has a way with him . . . a young David. Mind you, I reckon our Sally's taken to him too. . . .' 'Has t' only just noticed that? . . .' retorted Martha. 'Tha mustn't laugh now! The lass hasn't been well, and she's young for her age . . . a bit mollycoddled. And to see her face tonight! It's good for her to have a bit of life . . . but I'll keep an eye on her. Don't you worry!' 'Right,' said Simon, once again appreciating his wife's way of managing, 'but I've got to like that lad.'

The reedy organ died slowly away. The wonderful horses eased to a creaking walk and Josh and Sally crept quietly back to them. Simon took out his watch. 'Well, you two,' he said, 'it's getting late. We'd best be on our way.' Their faces fell, so Simon added, 'There's cattle to see to, and a long road ahead.' Josh remembered, in sudden fear, his dark home and the corpse under the stairs. His thoughts must have been mirrored in his face, for Simon added, 'Don't take on, lad. We'll see thee home.'

XXIII

ESCAPE FROM ROGER'S YARD

The sound of their footsteps changed tone, no longer ringing echoes from pavement on to shop fronts, but reflecting a muffled beat from the earthen surface of Roger's Yard. Darkness sat heavily on the close, foetid air, a darkness emphasized by the light sky high above, as the moon rose, invisible, beyond the rooftops. The Stockdales shuffled cautiously behind Josh, and Sally's eyes grew wider as they passed the first few open doorways and smelt the sweet, dirty smell of penned-in folk, so different from the clean smells of the farm. Ahead of them Josh's brisk walk began to drag as he brought them near his home.

There was the open doorway, framed in stone and pouring out a confused sound of raised voices into the yard. They waited in the darkness and saw a dusky red glow from the ashes of a fire and the cell-like room half-revealed by one yellow candle burning shakily with a thin, high flame. A box of a room, cut into by the smaller box for stairs, with no space for concealment, so that the blanket-shrouded form that was wedged under the staircase loomed enormously large, ominously present. Not that it had any shape, being merely a sack-like thing, save for the pair of grey old clogs that were thrust out from underneath and the knuckle of one skinny ankle that had escaped concealment.

One of the living folk in the room had begun to shout. The blousy, grubby, wet-lipped woman who must be Josh's mother was yelling incoherently. '. . . Th' old drape cow . . . it's good riddance . . . an' nowt less. . . .' Her head swayed uneasily from side to side.

This was more than Martha's sense of propriety and concern for her grand-daughter's upbringing could accept. She had known Roger's Yard was a poor place, though not that the *Craven Herald* had called it 'the natural abode of dirt and squalor', or she would never have entered the grimy slit that opened off Sheep Street. Taking the lad home had been a mistake.

Josh sensed some of this, though their surroundings seemed normal enough to him. His mother now sat collapsed in a corner, like a fat, bulging sack. Bert Ridding came downstairs after changing out of his best jacket and, with open shirt-collar to ease his breathlessness, carried a chair over towards the door intending to have a pipe. Some sound in the darkness of the yard caught his attention. 'Hello,' he said, 'what's this outside?'

Josh, badly discomfited by the scene his home presented, and still unhappy to sleep above the corpse, had been shuffling his clogs against the wall. He stepped into the candlelight and answered miserably, 'It's me, Mr Ridding.'

'Aye, Josh, so you're back then?' Josh nodded. 'I've told yer Mam I'll see to t' coffin and burial, though I doubt if she's taken it in. Now, hark to what I say!' He took out his pipe and began to fill it. 'I've settled in my mind today, while I were thinking things over, to give notice here, and move. . . . Now don't say owt, though I'll miss thee. We had some good readings together, thee and me. It's time I moved on and set up for miself. . . . You'll be master of this house now, lad.' Josh looked round and scowled. 'You'll be bringing in t' brass. I know you won't go her way,' Bert continued and pointed to the abject Florrie. 'Keep yer pride. . . . I know yer will. Many a night we've talked, and not unprofitably I reckon. You've been like a . . .' he looked at Florrie, then at the corpse, '. . . a nephew to me,' he added lamely. 'If you like, canst' come to t' Rechabites . . .? I'll take thee . . .'

Simon was fidgeting to be off and Bert heard him. 'What's this,' he said roughly, embarrassed to find his fatherly counsel overheard, 'hast brought young Whinsby to watch?' 'Nay Mister Ridding, it's only some folk I met at t' Show, they're good folk an all.' Bert spoke sharply into the darkness. 'Well, come inside, if you've seen all, and let us see you.'

Martha, her voice rising in scale with her sense of propriety and a sudden realization of the odd situation they were in, clasped Sally defensively and replied, 'No, we must be going.' 'Aye,' agreed Simon, and took a step back along the yard, 'let's get home.' 'But we've got to find Joady,' cried Sally. Josh darted forward at once, eager both to help and escape, afraid to lose his friends, who were fast disappearing down the yard. Bert followed, taking up the candle, and held it out of the doorway, calling into the darkness.

'Hey! Hey there! Don't rush off. I don't know who you are, but if you've owt to find in Skipton, then Josh's your man.' The footsteps halted. Bert followed. 'Don't run away!' he called. 'We're not running,' Martha answered defiantly, and stepped forward once more. Bert took in her respectable dress and honest face with one shrewd glance. 'Wait on,' he said in a low voice, 'I'll see thee to t' street.' As they walked in single file, the candlelight falling unevenly on black walls and stained earth, he confided rapidly in her ear. ''Tweren't a pretty sight in there, not for ladies, but don't blame the lad . . . and don't judge him by that woman. Reckon I only stayed on for his sake. You get fond on him.'

Martha shuddered at the thought of Josh's dishevelled mother. Josh had seemed a nice clean lad, if a little tattered in his clothes, but then, lads would be lads. 'Don't judge harshly,' added Bert, as if echoing her own thoughts, and Martha, safe now in the lamplight, could relax in recognition of the honest sincerity in Bert Ridding's voice. It was a voice of the kind of folk she knew and trusted.

Bert took heart at this and called down the yard, 'Josh, come here!' Josh came running, to burst out into the light like a ragged ghost. 'Go along, lad, and help these folk find their friend.' The lad's face flushed with such a beaming smile. He gave such a look of heartfelt gratitude to Bert Ridding, that

Martha offered an arm to him almost as easily as she would to her own grandson.

'Is the lad coming?' asked Simon. 'Aye,' she answered, 'he'll do.' She indicated the yard entrance with her head and whispered, 'We mustn't judge too hasty . . .' 'Judge not, that ye be not judged,' agreed Simon sententiously as he led the way. 'We'd better look in t' pubs,' he announced with a sigh.

It was Josh who found him, in the bar of the Red Lion. The room was filled with guttering candlelight and smoke, while on the bar the tray of tripe morsels had become a sludge of vinegar and water mixed with cigar butts and pipe scrapings. A fat-bellied man with rolled-up sleeves snored close by the fire, his mouth open and glistening with gold. His torso almost hid the lanky form wedged upright beside him, the round, weatherbeaten face tilted back against the wall, a smoothly worn thumbstick held in slack fingers. 'Yes, that's him,' growled Simon as Josh dragged him across the room. 'Come on, we'd best get him moving.'

It was easier said than done. Joady's legs seemed to have a life of their own, at one moment stiffly fighting every move, so that he threatened to topple over like an unstable statue, and Josh winked apprehensively at the looming mass of cowman above him. The next moment his limbs would turn to rubber, knees sagging as he lurched floppily over the shoulders of his supporters. When they were outside Simon sighed, propped his useless cow-hand up temporarily against a shopfront and wiped his forehead. 'We'd best run him like a sheep,' he said. 'How's that, Mr Stockdale?' 'I'll take his scruff,' and he seized the loose cloth behind Joady's collar. 'Thou mun take his rump . . . aye, grab his backside, lad!' Josh obediently took firm hold of the shiny breeches. 'Now, let him run . . . Whoa there!' Propelled from behind, held precariously upright, his legs working mechanically in response to the continued threat of falling forward, Joady was steered across High Street at a shaky trot. They fetched up in the stable yard of the Black Horse, where Simon up-ended him ruthlessly into the cart, wiping his hands on his jacket to signal the conclusion of the matter.

Josh, a little out of breath, ran round to help Martha and

Sally climb on to the seat and then took station by the horse's head as Simon buckled on the harness. He felt small now, alone and sad as his new friends from the wider world busily prepared to leave him. Save for the shuffle of post-horses in their stalls, the rustle of hay and a regular drip of water from the pump, the yard was silent. Josh rubbed his nose against the horse's cheek for comfort.

A heavy hand fell on his shoulder. 'All done, lad,' said Simon. 'Aye, Mr Stockdale.' 'Well, goodbye.' 'Goodbye, Mr Stockdale.' Josh bit his lip and looked the horse straight in its one uncomprehending, hazel-rimmed eye.

'Right,' said Simon as he climbed up the wheel. The whip cracked. The horse blinked and moved one leg forward. The whip cracked again and the cart rolled on towards the arched opening to High Street. Josh ran alongside until their speed increased and he dropped behind. Sally waved goodbye from her cocoon of shawl and grandmother's arms as he stood, watching, in the middle of the roadway until they were out of sight.

XXIV

PIPE-DRAWING

Josh was subdued and thoughtful when he returned home and crept up to bed. Friends made and lost were something to both celebrate and mourn. He would look out for the Stockdales again. Sally hadn't been bad fun for a girl. He wriggled his toes and grinned to himself.

There were also more serious questions to consider and he could in no way avoid them. He would have to be master of the house, there was no choice. His gran was gone, not that it worried him. Old folk died, and she was very old. But Mr Ridding was leaving, and that was much more serious.

To cap it all, the real meat of the day, he had his discovery of machinery and Mr Morrison to ponder and even a dream of escaping from the inevitability of the mill. Not that he could really make decisions himself. It wasn't expected of a lad and anyway his mind kept confusing him under a torrent of images, games of black plum, the elephant, Sally on the steamboat, the rampaging bull, Gran's dead hand. How could he hope to concentrate? He lay restlessly and waited for Bert Ridding to come up. Pipe smoke drifted through the window.

Bert himself had been musing. He felt stung into action by a small jolt, something that had oddly matched his mood. It was merely a chance remark from one of his old mates, a little joke about Bert, the soft romantic, and suddenly he realized how much he had let his life slip here in Roger's Yard. Pride and respect demanded that he should act at once, and he had at least begun. He stretched, knocked out his pipe, and went upstairs.

'Hello, Mr Ridding.' 'Goodnight, Josh.' 'Mr Ridding?' 'Aye.' 'Why are you going?' 'It's time I did. There's t' future to think on, an' I've things to do. Now don't go on about it. Tha's heard my story.'

'Mr Ridding, I've a lot of things I want t' ask yer . . .' Bert was bent over his tin chest, packing away his clothes. 'Right,' he grunted, 'wait till I'm settled. Then go ahead.' Josh waited. Bert got into bed. His breathing became even again. 'Mr Ridding, do I have to start as a half-timer and go to t' mill?' 'No, lad, there is no slavery in this country. But you might go hungry wi'out work.' 'I didn't mean that at all. Any road they only pay yer a penny an hour when yer a half-timer.'

'That's true enough, so what's on yer mind, young man?' 'I had words wi' Mr Walter Morrison this aft.' 'Didst'a now? Words wi' the millionaire indeed.' 'Aye, I were helping them wi' tractor. He said if I were keen I could go to see him about learning.' 'Did he now? That's not to be sneezed at. And are t' keen to learn?' 'Oh, aye. It'd be right grand. All that fun wi' t' motors an' outside an' all.' 'They'd fasten thee for seven year, but it'd be a skilled trade.'

'Should I go to see Mr Morrison?' 'Do you want to?' 'Aye!' 'He lives up at Malham, tha knows.' 'I can walk it.' 'What'll you say to him?' 'I don't know. I don't know at all. Reckon I'll try and find that engineer. Sam they called him, an' get him to help,' 'And he will,' said Bert to himself, knowing Josh's ways. 'But I'm going to do it, Mr Ridding. I'll go up there tomorrow. First thing.' Josh punched his pillow with both fists.

Then he rolled over to look closely at Bert. 'I were thinking about thy story.' 'Nay, I hoped tha hadn't heard,' said Bert. 'Dost understand? My mind were running on from what I were thinking before, how one day can change things so . . .' 'That it can, ' agreed Josh fervently as he thought over his Show day, 'but I never got to hear that tale about t' lead and tha' accident.' 'I can see I'll have to tell thee, or neither of us'll get any sleep.' Bert edged himself up his pillows and cleared his throat. Josh curled up happily by the wall. There was a limit to the amount of worrying he could do. Now was a time for enjoyment. He waited for Bert to begin.

'You know I went to work at Fell's after Grassington Moor mines closed down. Well, a few years back we got a big hydraulic pipe-drawing machine in t' works. It goes right up nearly to the rafters and it works off pressure from t' donkey engine. We use it to draw out gas pipes from refined lead.' Josh nodded to himself. He had kept watch on Fell's Leadworks from the towpath now and then in hopes of seeing his friend and, interested by the processes, had lingered until driven away. He had seen them firing the big crucible of scrap lead out in the yard, hurling dirty lumps of old drains, gutters, flashings into the dissolving heap, jumping hastily backwards, coat flaps over face, as a pocket of trapped water suddenly exploded in the heat and blasted out a spray of boiling steam and droplets of lead. Then they had tapped the crucible, driving an iron spike up the spigot hole, leaping back again as the red hot stream gushed suddenly out, hissing into the sand, and running it off into moulds until at last the shining pigs, oblong, with wedge-shaped handles, could be stacked on to an iron-wheeled trolley.

But Josh had never been inside the cavernous high-beamed shed, its walls slippery with lead-dust, where Bert worked in the light of mantel-less gas jets with their fish-tail spray of flame. The re-melting crucible gave out a thick smell of grease and tallow flux as the lead was softened ready for the hydraulic pipe press.

'It were Broken Hill Special, not old lead,' said Bert, 'and after we'd skived t' dross off the top we ran it into t' pipe drawer and left it to cool. You had to wait till it were just right, not so soft it ran out of shape, not too hard to draw. You wanted it smooth as butter before you put on the pressure. I were at the top, taking new pipe as it came out and running it to coils against the wall. Well, I just bent over to take one more length when all went bang! Like a cannonball hitting me it felt, for I were looking down into t' tube, and it shot me right across t' floor.

'What it were, was a blow-out of lead under pressure. It

happens now and then, you see, that lead cools unevenly and so you get a pocket that's left liquid. There, when t' pressure reaches it, bang – up it goes! The beams above that machine are sprayed with lead.

'Well, all the hands, all me mates, gathered round me. I shook meself and felt this stinging on me neck, so I put up me hand and a plate of lead fell away. There were a nasty smell, like burnt meat and I fainted, right sudden.' Josh thought with disgust of the hot lead hissing into damp sand.

'They threw some water on me. Work were stopped and old Mr Fell came into t' shop. 'What's going on here?' he said. 'What's happened to t' work?' 'It's Bert Ridding,' said somebody, 'there's been an accident. He's burnt a hole in his neck.' 'Well,' he answered right away, 'get him out of here, then. He's no use to me with holes in him.'

Bert chuckled quietly. 'But they gave me two weeks off wi' full pay. He'd always roar at you first, would Mr Fell. But you got to know him. . . .' Bert looked down at Josh. His eyes had closed at last. He was breathing peacefully. The candle was guttering to a liquid death. Bert blew it out. All was dark and silent in Roger's Yard.

XXV

MOONLIGHT

From black to white, moving through house shadows into the dazzling brightness of the limestone roadway, then back again into shadow, the Stockdales had rolled out of Skipton. The heavy masses of Church and Castle vanished behind them. Soon the sooty smell of town streets, the dying buzz of noise, even the glow of yellow cast by the fair were lost behind them. They were out on the quiet road and their own moving image was etched beside them by the rising moon.

The great day was almost over. Mr Barrett, having entertained a select group of gentry with dinner at home, had seen the Castle securely locked, received the urgent reports of his staff on the consequences of the Show to the property in his care and was now winding up his watch by the bedside as he looked out over the town. Walter Morrison had been driven home to Malham and had gone to bed early with instructions to be wakened at six, for he fancied a tramp over the fells before church. He had a number of ideas to think over.

The manufacturing families, whether old established and Anglican, or newly successful and still Nonconformist, were mostly asleep. 'Early to bed and early to rise,' was their motto. There were a few exceptions to this rule. Mr Scott, oblivious of the brewery and his debts, sat by an empty grate and swallowed brandy with a lugubrious face. Mrs Scott had already retired with a book, remarking as she passed the door of the drawing-room, 'I do wish you'd be a little more diverting, dear. After all, this is a holiday.' Of course John Bonny Dewhurst was still very much awake, scheming, planning for the future from the smoking-room as he stretched and tapped his bony fingers. After all, he had his eighties to provide for, and who could predict what the new century might bring.

Sleep was in order for all the sober, hard-working people of the town, for the Tawneys, Fred Manby, his tired eyes still sore from the day's glare, Baldisaro Porri, the shopkeepers, craftsmen, skilled hands. Others were less responsible. The navvies in Charlie's doss-house, having been ejected with considerable noise from their favourite inns, were now drinking at home. They had been singing for some time and clapping in rhythm as the bear cub tottered between them. Gianpiero, having made a little money at the fair, was deep in exhausted sleep. The bear had been liberally supplied with ale and by now was as incapable as his audience. The rest of Union Square ignored the row. They were used to it.

The canal yard and offices were bolted and barred. Old Bateman was snoring at home. Moonlight shone on the surface of the water, breaking into a pattern of rings where the ripples spread away from Badger's boat. He had Izzie cabined at last and as he grunted away the whole boat rocked slowly in response.

The Stockdales had been travelling silently for well over an hour, absorbed by the memories of that eventful day.

Martha had been thinking over all that had happened and now felt compelled to break the silence. 'What're you going to do about Joady?' she asked. 'Nay,' replied Simon, allowing himself time to ponder, 'nay I've given it little thought. But he's a good cowman. I'll tell thee what though . . . He's not going to t' Show again. Skipton's like Babylon to Joady.' Silence followed as they passed a farm and the irritated clucking of a disturbed hen followed them down the road. 'I think I'll let him sweat for a day or two . . . be a bit short with him. It'll do him good . . . give him a lesson.' Martha smiled to herself, knowing how long this stern Simon would last once he was at home. 'Perhaps you should,' she agreed.

Joady opened his eyes. A clear round face smiled down on him as a mother does above her child's cradle. Joady smiled back at the moon as he slowly recovered. He felt planks beneath his body, heard the sound of hooves, closed his eyes and was beginning to doze when he heard his own name spoken. Simon's unconvincing threat to be 'short' with him rang in his head with bell-like clarity. Joady smiled at the moon once more. A repentant sinner, happily accepting his punishment, he thought of his loft above the cattle, the scent of hay, the stirring of the gentle beasts below him, and so dropped away into a forgiving sleep.

The silence lasted for some miles, until Sally stirred from her grandmother's lap and asked in a tired, thoughtful voice, 'Grandma?' Martha shook out of her slumber. 'Aye, lass . . . why aren't you asleep?' 'Grandma . . . I wish Jasper was coming back with us.' 'Well, lass . . . they have to go, you know, there's nowt lasts forever. Now, settle down, close your eyes. We've still a good hour's ride ahead.'

The steady motion of the walking horse filled them all with a sleepy contentment. Dozing lightly, so it seemed, they came at last in reach of home, the high dale with its narrow ways bordered by stone walls and rough pastures. Silver fells rose above them, the crust of old lead workings, like distant molehills and runs, topping Grassington Moor, whilst Malham Moor was a smooth sweep of shadow, concealing the dark waters of the hidden tarn beyond.

Soon came the first glimpse of the shining roofs of High Dubb Farm, folded around by the hills. Gleaming white stones of the field walls were edged by deep borders of shadow. As they passed close by the knotted hawthorn trees that sheltered the lane, sprays of dew fell on their faces.

Simon, drowsing as he drove, shook himself and sat upright. Martha shivered in sympathy and, opening her eyes, drew Sally closer into the shawl. 'Home again,' said Simon, 'praise the Lord in every place.' 'Aye,' replied Martha, 'and what a day!' Simon began laughing quietly to himself. 'What is it, Simon?' 'Nay it's nowt . . . I were just thinking to myself "Thank the Lord for t' butter."' The horse drew up of its own accord. Simon climbed stiffly down, stretched out his arms as if in greeting to the moon, brought the key from under a stone and unlocked the farm-house door. 'Well, I'd best see to t' beasts,' he said, and began to walk across to the shippon.